Internal Organs Create Your Heart and Mind

Life Memory of 500 Million Years
Mind Represented by Internal Organs

– Katsunari Nishihara –

Translated by Isao Hirano

UPFRONT PUBLISHING
PETERBOROUGH

Katsunari Nishihara

www.fast-print.net/store.php

INTERNAL ORGANS CREATE YOUR HEART AND MIND

ISBN 978-184426-869-6

First published 2012 by
UPFRONT PUBLISHING
Peterborough, England.

An environmentally friendly book printed and bound in England by
www.printondemand-worldwide.com

FSC
Mixed Sources
Product group from well-managed forests, and other controlled sources
www.fsc.org Cert no. TT-COC-002641
© 1996 Forest Stewardship Council

PEFC
PEFC/16-33-415
PEFC Certified
This product is from sustainably managed forests and controlled sources
www.pefc.org

This book is made entirely of chain-of-custody materials

Contents

Preface

This book's title is the answer to research into the whereabouts of the heart and mind. This research theme is the most difficult problem among life sciences, and nobody has yet showed a clear-cut answer by modern scientific methods, because what in the world is the mind and the spirit has not been fully grasped scientifically. In order to solve what these are and where they dwell, above all, you have to understand correctly what life is. Furthermore, so long as you don't understand how the evolution of the vertebrates including mammals, to which human beings belong, has occurred, you cannot find the whereabouts of the mind and the spirit. The reason is that it is the ironclad rule of morphology to search for the origin of them, if you want to know the essence of the organs, since the animal organs have various functions. The study to search for the origin is phylogeny, which means tracing back the process of evolution.

The mind and the spirit must be functions of specific organs, or put another way, the work of them. They must be the same substance without mass as body temperature, voice, respiratory movement, or muscle movement. However, it is also very difficult to verify the whereabouts of the mind using the other animals except human beings. Only human beings, Homo Sapiens, among the mammals that are the most distinguished family among the vertebrates in the animal kingdom, can demonstrate the mind or thought. On top of that, human beings can show or express that there exists the mind or thought using words or symbolic characters. The search for the whereabouts of the mind is no less difficult than to elucidate the reasons for the development of intractable immune diseases which occur only in human beings by the structural defects of the body peculiar to human beings.

The mind is the function that the organs, tissues and cells of animals possess. Since the starting point of multi-cellular animals is a protozoan consisting of one cell, the origin of the mind is energy generated by the work of cells that are the minimum units of life. The immune diseases peculiar to human beings, to think profoundly, are ones caused by the deterioration of work of organs, tissues or cells, by any energy. That's why you can't locate the whereabouts of the mind, unless you can elucidate the enigma of the development of immune diseases that occur by the change of energy metabolism.

The author is a doctor of stomatology who has been a clinician for 37 years. The 20th century was the age in which 'a stomatologist called a superior doctor' has been expelled by 'dentistry' representing organ-oriented medicine of the United States of America. But, in Japan, there existed a medical scientist named Tooru Shimamine, who had been endeavouring with a dignified attitude in order to revive a doctor in the Oral Cavity Department in the forefront in the world from the Taisho period until World War II of the Showa period. This plan to establish a Medical University of the stomatology was obliged to become extinct by the demise of Dr. Shimamine in the 20th year of the Showa period.

The author entered Tokyo Medical and Dental University more than fifty years ago. Although I chose this course in vague anticipation of studying medicine about the face and the oral cavity, I lost my lofty ambition at the period of the technical training course, and was quite disappointed around graduation. But, as I had still received the lecture on general morphology in anatomy by Shigeo Miki, I hoped to research the evolution of the vertebrate in a full scale once in my lifetime. After two years as a dental intern, I entered the postgraduate course of faculty of medicine of Tokyo University, which was Shigeo Miki's alma mater, and I took steps forward researching medicine. I noticed it was the best way to become a proficient stomatologist whom Tooru Shimamine had proposed, and become a scholar of medicine such as Haeckel, Roux, Shigeo Miki, Helmholz, Herz or Young. Though I may appear to praise myself, I thus aimed at becoming a stomatologist. A proficient stomatologist refers to a doctor who always assumes an attitude to treat by medicine based on the evolution of vertebrates.

Around the time I entered the postgraduate school was just the start of campus strife originated from the faculty of medicine, Tokyo University. I got involved in this campus strife, and I was obliged to deviate from a scholar course. I received the degree of Ph.D. in medicine by the thesis. Since the time when I received a lecture of general morphology by Shigeo Miki in my school day around in the

36th year of Showa, I was very nearly convinced that three enigmas of the phylum will be elucidated, if bones to define vertebrates can be synthesized artificially. As organs characteristic of this phylum are the branches of intestinal respiration and the lungs, I have believed that deep research into artificial bones and external respiratory organs would lead to the solution of the three enigmas: the causal factors of evolution, the riddle of the immune system, and the development of bone marrow haemopoiesis.

When hydroxy-apatite – the mineral substance of bone – was synthesized in our country, I started studying an artificial dental root using it. I succeeded in the development of a tooth of gomphosis analogous to a human tooth (with the periodontal membrane and the cementum) just one year later. This artificial dental root has not been developed for 150 years in America in spite of earnest research. In the following three years, I succeeded in becoming the first person in the world to induce myelopoitic nests in muscles and osteoblasts from muscle cells, using an artificial bone marrow apparatus made from porous apatite. By using biomechanical stimuli nobody had thought of, namely energy caused by the hydrodynamic flow of body fluid occurring by muscular movements, I succeeded in developing the methods to induce cementoblasts, haemopoietic blood forming cells, and osteoblasts.

The author has doubted for a long time whether the evolution of living organisms can be explained by mutation and natural selection. Limiting living organisms to vertebrates, in their evolutionary process gravity turned out to have played a great role through the development of an artificial dental root and an artificial bone marrow. I have elucidated through my way of experiments and insight how organs or skeletons of vertebrates have evolved by experiencing gravity, and by summing it up I published 'Seibutsu wa Jyuuryoku ga Shinka saseta' [Gravity Made Living Organisms Evolve] (Koudansha Blue Backs Publishing).

Since it is the basis for this book, I will explain it by summing up in brief:

Tracing back the evolution of vertebrates reaches an archae-ascidian, the ancestor of an ascidia of protochordates. An ascidian, as you know, absorbs nutrients in seawater, sticking to the seabed and swaying in the movements of waves. An ascidian is composed of the intestine to absorb nutrients and the branchial intestine for respiration. It is a living organism like a sac that takes in food and seawater containing oxygen from the mouth, and removes wastes from the gill slits and the anus.

However, the young of an ascidian has a fish-shaped figure having a tale like a tadpole, and does not adhere to the rock at first. A conventional theory of the evolution formula about vertebrates has showed that the change to a fish-shaped figure occurs if a living organism becomes an imago, remaining in the larval state. The author has succeeded in neoteny (gaining sexual maturity, although remaining in the larval state) of an ascidian by experiments, and has searched for the causal factors of the evolution of vertebrates. But, through the study of the evolution of the backbone, an ascidian turned out to be monosegmental, and neoteny turned out to be a blind alley. And it has been elucidated that there existed a stage of the primitive revolution called multi-segmentation of the ascidian. Then, I have elucidated the development of polarity of the mouth - the anus based on gravity. Multisegmental ascidians started swimming in the seawater are subject to the law of inertia based on gravitational force by cephalad movement, by which organs such as the intestine aligned even to segments move backward one after another by the law of inertia. The body elongates, and the intestine polarizes into the mouth side and the anal one to change its form. Thus, the evolution of vertebrates conceivably proceeded to the next stage, and finally moved into new stages through landing drama (terrestrialisation).

In the reproductive experiment of landing drama occurred in the vertebrates 300 million years ago, I actually made a shark land to verify the change of the constitution into land vertebrates. That is to say, by landing a shark that performs gill respiration with quite low blood pressure, blood pressure of the shark rises by writhing in agony to be able to perform air-respiration by gills. The increase in blood flow and rise in blood pressure brings about the increase of streaming potentials, by which the trigger of the genes of mesenchymal cells is pulled, cartilaginous bone ossifies, and at last blood forming tissues develop in the bone marrow. Thus, each evolutionary stage of the vertebrate has not occurred by mutation, but occurred by energies such as gravity or streaming potentials, and expression of the genes of cells by chemical substances. For example, ossification of cartilaginous bone has brought about myelopoietic nests. This is the verification of the Use and Disuse theory that shapes of the body change with behavioural patterns transmitted from generation to generation.

Research into the whereabouts of the mind found an unexpected breakthrough. Claire Sylvia afflicted with primary pulmonary hypertension underwent the most advanced simultaneous heart and lungs transplant operation in those days that developed internal organs transplant by Alexis Carrel in 1988, and her writing 'A Change of Heart' was published in 1998. In the novel, the mind of Claire Sylvia has been reportedly changed to the mind of a young donor. Just when I have clarified through research into the evolution of phylogeny using sharks that the mind dwells in the intestinal visceral system originated from the branchial apparatus in which the muscles of the face, the tongue muscle, and the heart and lungs are unified, I knew Clare Sylvia's private papers. Besides, since the author has discovered that the lungs and thymus of the vertebrate alone get into the pericardial cavity existing in the surrounding of the heart, the simultaneous heart and lungs transplant is understandably easier than that of the heart alone, and what's more, the simultaneous transplant including the tongue will be more advantageous, because as far as the diaphragm is the part changed from the shark's immovable tongue. Thus, as the result of the example that might be called 'the human body experiment', the whereabouts of the mind finally tuned out to be located demonstratively.

In this book, research on evolutionary study, immunology, and the development of myelopoiesis, have all been systematized based on the same law. This law refers to 'the principle of the conservation of energy' that is the greatest scientific achievement in the 20th century. Until now, life sciences have been constituted by 'the law of conservation of mass' of the 18th century all over the world. This book gives you elucidation and explanation of life sciences by accepting energy as substance without mass. This will make the discussion of the mind, spirit, and thought available, for they are all the same energy as body temperature, respiration, or perspiration. The way to study by accepting only materials with mass might be real materialism. Accordingly, since this book has a quite new idea on its foundation, you may find repeated writings to some extent. I hope the new idea will sink deep in your brain by repetition.

In publishing this book, I was too much obliged to both Ichizo Tsoji and Mariko Hikata of NHK publishing, to whom I am very thankful from the bottom of my heart.

July, 2002 Katsunari Nishihara

Internal Organs Create Your Heart and Mind

Life Memory of 500 Million Years
Mind Represented by Internal Organs

– Katsunari Nishihara –

Translated by Isao Hirano

Chapter One

The Mind Changes by the Simultaneous Transplant of the Heart and Lungs The Mind is Energy

Can the methods of contemporary life sciences locate the whereabouts of the mind? Though a Canadian brain surgeon, Penfield Wilder, searched for the whereabouts of the mind through surgery of the brain, what he found was only the fact that the mind did not exist in the brain. Similar research has ended in failure for many scholars and must therefore have had some intrinsic defects.

There lived a French medical scientist named Alexis Carrel, a scholar who published L'Homme-cet Inconnu [Man, The Unknown]. He was awarded a Nobel Prize for physiology and medicine for the new method of angiorrhaphy and organ transplantation developed in 1912, almost a century ago at Rockefeller Institute. What is especially noteworthy in his book is the following: 'whenever a man communicates with another man, no matter whether it may be spiritual, the relationship between husband and wife, between mother and child, or between father and child, it is connected materially beyond time and space.' That is to say, he says that the spirit or mind is connected from the past to the future beyond distance. This is the conclusion of the research that applied 'the principle of the conservation of energy', the greatest achievement in science in the 19th century, to life sciences.

Any phenomenon called the mind, spirit, or soul, is in fact mass-less energy. The theory that time and space are in relativity through the intermediation of light, mass-less energy, is Einstein's 'Theory of Relativity'. By this view, a universal theory of 'the law of the conservation of mass' in the 18th century has been replaced by 'the principle of the conservation of energy'.

Though it seems to be a little difficult to understand, you have got to understand by all means the outlook in the universe of the 20th century in order to search for the whereabouts of the mind or spirit – mass-less energy becomes equal to the material having mass in an extreme situation (for example, in a nuclear reactor), because any of the mind, spirit, thought, or feeling, is the same mass-less energy as body temperature. Life energy, just like temperature, respiration, perspiration, voice, supersonic waves, or biophotogenesis, is the electric phenomenon in the body, which is the vortical rotation of electrons passing concurrent with metabolism (remodelling) that is essential for life activities. In the 20th century, the outlook in the universe based on 'the principle of the conservation of energy' has not been ubiquitous. If you understood this fact really, your way of living and religion would change.

The Research Method to Search for Whereabouts of the Mind

Once you recognize the mind to be energy, you will easily find the method to search for its whereabouts. In the same way as in studying hormone secretory organs, removing or destroying the organs that are estimated to be the whereabouts of the mind, appears to solve the problem. In reality, there was the age of ordeals for animals: many dogs, cats and monkeys were sacrificed to have their brain cortex completely scraped off, to have their brain stem cut, or to have lobotomies to cut the frontal lobe of the cerebrum undergone. However, the result was the fact that you could not find the whereabouts of the mind, no matter how you tinkered with the brain.

The starting study of self and not-self immunology was the experiment by a French female scholar, Le Douarin Nicole, to produce a chimera of a chicken and a quail in the embryonic stage before hatching. A foetus before hatching is of immunologic tolerance, or latent with the mechanisms of histo-immunity, even in the higher vertebrate. Since this is the same situation with mammals, it is possible to perform exchange transplant of tissues at this stage.

Le Douarin carried out the following exchange transplant: when each fertilized egg of a quail and a chicken cleaved to become a neurula, she carried out the exchange transplant by cutting each neural crest that was to become the brain or spinal cord. With it staying alive, a chimera of an quail and a chicken, namely, a chicken with the brain and feathers of an quail was hatched. Does the chicken with the brain of a quail cry like a quail?

The author has succeeded in transplanting the brain of a myxin of cyclostomes and the brain of a shark into the brain of 20 newts and 10 rats; the details will be described later, but the behavioural patterns of a rat and a newt did not change at all and they survived for five to six months, safe and sound. Namely, the nerve cells of the brain is just a circuit of electrodes even after transplant, and does not generate specific electric currents that express like or dislike. A chicken having the brain of a quail cries, of course, like a cock. Namely, as Penfield states, the whereabouts of the mind does not dwell in the brain. It is conceivable from the beginning that experiments using dogs or monkeys are not sufficient for the study of locating the mind. Differing from hormones, temperature and respiration, even though it is a kind of energy, the mind is of such the highest order that it can be expressed only by speaking language. There are studies on making apes talk, using symbols; however, it is quite difficult to discuss the mind or thought delicately. But, surprisingly enough, the research to what may be called 'human body experiment' was already carried out in the United States of America, and its application to a patient has been reported, which verified the whereabouts of the mind by human organ transplant predicted by Carrel. It is the book named A Change of Heart: A Memoir.

This book makes it clear, based on the result of a human heart and lungs transplant that actually happened in the USA, that the mind exists in the heart or lungs, namely in the intestinal visceral system, and furthermore, verifies by tracing back ancient events that the mind exists in the visceral intestinal system. I shall verify this by consulting the research of the evolution of vertebrates, using phylogeny, comparative anatomy, and Experimental Evolutionary Study that I devised.

From 'A Change of Heart: A Memoir' [Claire Sylvia with William Novak, Forward by Bernie Siegel, M.D. Little Brown and Company Boston, New York, Toronto, London]

The heroine of this book, Claire had suffered from an intractable disease and underwent the heart-lung transplant several years earlier. Before long, she came to think that the organs of the young man who died of a bike accident lived in her body accompanied by their own consciousness. A recommender of this book, Dr. Chopra's view of life and death is not compatible with ours, because the death of a man in the Christian world, that is to say, the death of the brain is not the death of cells, but is something like an agreement between men as well as a law. Besides, it is not Natural Law but an artificial administrative law. As Claire's book is a precious record about the whereabouts of the mind, let's look around quoting from her book. The following sequential quotes are from her book:

"Before long, I began to feel that I had received more than just new body parts. I began to wonder if my transplanted heart and lungs had somehow arrived with some of their own inclinations and memories. I had dreams and experienced changes that seemed to suggest that some aspects of my donor's spirit and personality now existed within me"

(From Chapter 1).

She began to feel that the heart did not only take a sole role of the pump, but differing from other organs, had any connection with emotion. She began to notice that the term 'heart' was often used in figurative expressions: 'broken hearted', 'take heart', 'lose heart', 'wear one's heart on one's sleeve', 'heartless', 'pure heart', and so forth.

"Even the most conservative cardiologist will acknowledge that the health and functioning of the heart are affected by certain emotional realities, including loneliness, depression, and alienation. And while it is commonly accepted that the mind and body are deeply connected, we don't have nearly as many images or phrases pertaining to, say, the liver, the pancreas, or even the brain"

(From Chapter 1).

One day, Claire had an unusually vivid dream as if it had been a reality.

"It's a warm summer day. I am standing in an open, outside place, a grassy field. With me is a young man who is tall, thin, and wiry, with sandy-coloured hair. His name is Tim, and I think his last name may be Leighton, but I'm not sure. I think of him as Tim L. We're in a playful relationship, and we're good friends. We kiss – and as we do I inhale him into me. It feels the deepest breath I have ever taken. And I know at that moment the two of us, Tim and I, will be together forever."

(From Chapter 1).

Claire attended a meeting of a group of people who had all undergone a heart-transplant, and knew that the other members had felt the change of their identity. Amongst them were Mario in his fifties, and Thomas in his forties, who both had had a young man's heart-transplant.

After the transplant, Mario and his wife noticed a number of changes in his habits. He had never liked bananas before; he now loved sweets.

And while he had previously been fastidiously neat, after the transplant he was far more relaxed.

"Thomas, who was in his forties, took on a completely new personality after the transplant. Before the operation, from all accounts, he had been shy and introverted. But when he joined our original support group a few months after his transplant, Thomas, who now wore a baseball cap all the time, had turned into a big, exuberant, talkative child, like a nine-year old boy in a man's body."

(From Chapter 11).

After that, Claire was astonished to know that the name of her actual young donor was the same as that of the young man in her dream, finding the obituary of a young man in a library. After difficulties, she searched out the family and came to be accepted as a member of the family. In a later chapter, the explanation by Dr. Deepak Chopra has been introduced, by whom the theory of cell memory has been applied widely.

"Rather than seeking a supernatural explanation for such incidents, one could see them as confirmation that our bodies are made of experiences transformed into physical expression. Because experience is something we incorporate, our cells have been instilled with our memories; thus, to receive someone else's cells is to receive their memories at the same time."

(From Chapter 18).

And also Claire learned about the work concerning memories of organs of Candace Pert, a biologist.

Human emotions, says Pert, are triggered by neuropeptides attaching themselves to receptors, which stimulate an electrical change in neurons. While peptides were known to exist in the brain, Pert and her colleagues have found them throughout the body, including the heart. There are even neuropeptides in our stomachs, which, I suppose, gives a whole new meaning to the expression 'gut feeling'.

(From Chapter 18).

'A Change of Heart' is the voice of the soul of Claire Sylvia whose mind has changed to that of the donor. It has been completely forgotten in modern medicine that the lungs are such a large organ equivalent to a liver originated from the intestinal tract, and are the source of life. A heart-transplant alone does not cause such a drastic change of the mind.

Preference in foods means, from the beginning, the inclination of absorption in epitherial cells of the intestinal membrane. Colour is determined by a specific wavelength of light. Light is an electromagnetic wave, a mass-less energy. The eye is the structure of conveying according to like or dislike of the intestinal tract the information to muscles, at the paraneuron (cells having both the nerve connecting nerve cell and its effective organ, and its effective function) attached to the intestine for respiration of the branchia. For living bodies, absorption of the substance having mass from the intestinal membrane is quite the same as absorption of mass-less energy from the apparatus attached to the intestine (the eye). For example, the former is oxygen or nutrients, and the latter is light.

A heart-and-lung transplant may be said to be the pinnacle of transplant medicine of the 20th century applied to the human being, in which angiorrhaphy and organ transplant method developed by Alexis Carrel in 1992 were undertaken sixty years later. While I will describe in detail in the later chapters the research by Experimental Evolutionary Study using a landed horned shark (Heterodontus), I disclosed by this study that only in mammals, the lungs and thymi enter into the pericardial cavity, and the diaphragm develops from the bottom of the pericardial cavity. That is to say, into the outer sac surrounding the heart enter the lungs and thymi. The mammal is the only animal, in which the tongue, heart, and lungs are a unit of the external respiratory system derived from the branchia. It is the method of a heart-and-lung transplant that has applied this fact concerning evolution implicitly. What is more important than anything else, is that even if the brain of a shark is transplanted to that of a rat, or even if the brain cells of a human foetus is transplanted to the brain of a human adult, brain cells supply electricity simply as an electrode just like a transistor. Accordingly, the transplant of brain cells does not change personality or the mind at all. On the contrary, the fact that transplant of the internal organs changes the mind or heart, shows that the whereabouts of the mind exists in the internal organs.

Unnoticed Record of the Soul

This record of Silvia's soul shows not only a splendid achievement of a heart-and-lung transplant that is cutting-edge in organ transplant medicine in the USA, but also is epoch-making in having verified the whereabouts of the mind. However, since the soul or mind, being the essence of life, has been proved to exist in the visceral intestinal system such as the lung or heart, this record of the soul has been ignored by contemporary medicine. Today's medicine in the world consists of immunology of self and not self. Owing to this immunology, organ transplant has come to be available. But, if the mind or soul exists in the internal organs, it turns out that real self, identity, is not Major Histocompatibility Complex (MHC) existing in the membrane of the leukocyte, but is inclination of digestion-absorption ability by the intestine, namely, preference. Though I describe more in details, life phenomenon is the system to overcome ageing, by remodelling old cells while allowing nutrients and oxygen digested and absorbed from the outer world to react, and with turning energy's eddy around. Remodelling being the essence of life begins at absorption of oxygen and nutrients from the intestine. The ability of digestion and absorption in the intestine, and preference are nothing but the source of desire. Namely, the ability of the intestine represents the mind. Therefore, transplant of the intestine leads to transplant of the donor's desire. Desire is the source of the mind. MHC that the membrane of the leukocyte has, which was actually clarified by the author's research, does not work mainly for discriminating self from not self, but does work as the structure to destroy those damaged cells and reutilize by finding out open seams of the membranes of the cells that are old and ready to break, or defective cells such as tumour cells. What is troublesome with American medicine above all, is that the condition of organ transplant being the death of the human being whose brain function stops, even if the heart is alive, will be called off, if the mind or soul being the essence of life exists in the internal organs. In fact, since life develops from the intestine, the man lives as long as the intestine lives. If this fact is revealed, as organ transplant commits homicide according to the law of nature, organ transplant could not be performed any longer.

Organs Attached to the Intestine

Namely, the soul of Claire Sylvia appeals: since more advanced life science will be able to let a donor come back alive, the internal organs should not be removed, as long as they are alive. As this perplexes medical scientists, this voice of the soul has been obliged not to prevail too widely.

Diaphragmatic respiration by the parasympathetic nerve is Japanese culture of 'gut', while voluntary respiration to breathe largely with the chest is that of the sympathetic nerve. Respiration by westerners having culture of 'head' rich in spirit shows predominance of voluntary respiration.

In America, no matter how vigorous the internal organs are, it's all over for you, when your brain is damaged and you cannot speak, through being unable to express your intentions. If it meets the conditions of brain death, your internal organs are soon to be removed. I heard even in such a case that can be recovered from a vegetative state by hypothermia therapy developed recently in Japan, the internal organs are removed by judging it to be brain death in America.

If a donor is an erotomaniac, the recipient will become an erotomaniac. Preference for colour, like/dislike for foods, and preference for sexual desire, are all preference for the ability of absorption and excretion in the intestine. The branch organs of the brain presiding over sight, smell, touch, hearing, and taste, are actually the attached apparatus for the intestine controlled by the branchial intestine (the intestine moulding the branchia).

Please remember that substance with mass is equivalent to energy without mass, that is, the same for life. They refer to Shiki (body) and Kuh (emptiness) in Buddhism. Shiki refers to substance with mass. Ku is equivalent to energy without mass that is light, space, time, and heat. As far as Buddhist senses - such as sight, hearing, smell, touch, heart and will - and Buddhist consciousness are concerned, Goun (five elements) - namely Shiki (body), Jyu (sense), So (imagination), Gyo (will), and Shiki (judgment) - has been adopted as the 'mind and body method' to grasp all things in the universe. 'Shiki=body' refers to body and substance with mass such as foods or oxygen. 'Jyu=sense' is work of cells, which is energy. 'So' refers to imagination that is energy without mass. 'Gyo' refers to muscle movements moved by function of the mind, which is work of muscle cells, namely, energy. 'Shiki=judgment' refers to consciousness, which is work of the brain cells connecting the outer world and the mind being function of the intestine, namely energy without mass. You will notice that the other senses except smell and taste among the five senses are all mass-less energy. This fact makes us realize how fragile materialistic thought in the 19th century that this world has consisted of substance with mass alone, has been.

Chapter Two

Where Does the Mind Come From? Explored by Phylogeny and Clinical Medicine Evolution of Life from the Intestinal Point of View

'What face is to mind, body is to spirit' is often said. This means that the face represents your heart, and the body with swelling muscles represents your mind. The face is called 'the facial cranium' or 'the visceral cranium', because the face with the oral cavity centred is really born from the internal organs of the branchial intestine. A jawbone called a heavy-jawed face was originally a cartilaginous bone of the branchia. The branchia is the structure for respiration, the apparatus for breathing at the age of fish. The author has clarified by a series of research that the ancestor of us had been really a horned shark belonging to the cartilaginous fish of the primordial vertebrate. In the visceral intestinal system having the apparatus to breathe, dwells in fact the mind. That's why 'the face and the mind' is often said.

The face is composed of visceral muscles of the branchial intestine presiding over respiration, the most important function in life. There are three kinds of intestine: besides the branchia, the intestine for digestion/absorption, and the intestine for superfluous nutrients and excretion for wastes (the excretory intestine named by the author). The soul, self, or mind dwells equally in these visceral intestinal systems. The genuine work of the intestine is to produce wandering blood cells (cells remaining in the oldest unicellular animal, such as erythrocyte, leukocyte and so on) to carry nutrients and information throughout the body.

Until modern times, in Japan, China, and Europe, a doctor of the oral cavity handling the face that is the pivot of life had been recognized as the best doctor of all. In the 20th century, this 'Stomatologist' was replaced by the dentist, based on organ-oriented medicine in America. The author has revived the idea of the 'Stomatologist' by himself and acknowledged himself to be an 'Stomatologist' about ten years ago. Thus, I have created Gravitational Evolutionary Study of the vertebrate though not perfectly, or have studied the development of the face, and the development of the mind and spirit.

Buson (a famous Japanese haiku poet) wrote,

'Moto o wasure sue o toru tsugiki no ume'
Forgetting the origin,
Choosing a tip,
A grafted plum tree. (version by the translator)

Medicine in Japan is exactly so. With unsuccessful grafting, the arrival of organ-oriented medicine from America into Japan has excited considerable confusion today. Even in organ-oriented medicine, organ transplants popular in America can be said to be an experiment for research using the human body. The actual heart-and-lung transplant being a kind of human experiment, has verified that the mind dwelled in the internal organs. The mind exists in the heart in part, but the real whereabouts of the mind is in the lungs for the most part, because the heart belongs to the vascular system of the branchia, and the lung is composed of the intestinal epithelium of the branchial intestine. In the whole of the intestine unified the nerves of the intestinal epithelium and muscles, dwells the mind. The heart can be considered to be part of the muscles of the vascular system of the intestinal epithelium of the lung having no muscles. In order to grasp correctly where the mind and spirit dwell, it is necessary to elucidate the law of evolution concerning the intestine based on the gravity-corresponding evolutionary law. On top of the research, it is necessary to substantiate it by a real man as an example.

The Brain And The Mind

Today, it is taken for granted in Japan that the mind exists in the brain. As a matter of fact, the author also considered, while he was a student that the mind existed in the brain. When talking with a German teacher, I was asked where the mind existed; when I said that naturally the mind existed in the brain, he mocked me with a little look of disgust. I remember that he had persuaded me by giving an example of Goethe's 'Faust', 'Herz exists in Herz of the chest, and Geist (spirit) exists in the brain. In those days, I had thought that theories of the mind existing in the heart was ignorant ancient people's knowledge, on the basis of which the organ was named. In those days, I did not know at all that ancient people had grasped correctly the truth with the thought by the cerebral limbic system (refers to the visceral brain, which exists in the centre of the brain in the primordial vertebrate, and in mammals is located at the edge of the brain). Since then, I have been thinking of the validity of the German's saying. Just around this time, I had been receiving an anatomical lecture of 'General Morphology of Organisms' by Prof. Shigeo Miki (associate professor in those days) at the university. I had no idea in those days that Miki's Morphology had been the first step to lifelong studies that have published 'Science of the Face', 'Gravity-corresponding Evolutionary Study of the Vertebrate', 'Medicine for the Face', 'Science for Health and Diseases', 'Clinical Phylogeny', 'Evolutionary Study for the Baby', and 'New Immunology'. The author, on reaching an unlucky age of forty- two, has been treating diseases of the whole body and seeking the causes of diseases, by the new way of thinking called 'medicine for the face', by recognizing the visceral cranium of the facial part among the human face as complex organs representing life. To my surprise, since I chose the course of a scholar at the age of twenty-six, all I did was nothing but a doctoral thesis even at the age of forty. As the result of thinking of achieving self-fulfilment effectively during the rest of my life, nothing but the new way of thinking occurred to me.

Fortunately, the face is the entrance for respiration most vital in life and food. On top of that, the face is equipped with the eye, ear, nose, tongue, and skin that recognize the outer world, and the apparatus for expressing ideas, and is said to represent personality or the mind. I had anticipated in finding the whereabouts of the mind through studying the face by any chance.

Development Of Artificial Organs

Five years after when I set up the environment to be able to lead a research life freely, continuing clinical research unobtrusively, I succeeded in developing a mammalian type of artificial dental root having the periodontal membrane, cementum, and alveolar bone proper, using ceramic artificial bone in 1988 for the first time in the world: a tooth as an artificial organ replacing the genuine tooth with gomphosis. The conventional dental implant is of the type adhesive to bone (ankylosis) and is not available for mastication, which is the system seen in reptiles in which they get new teeth after breakage (Fig. II-1.).

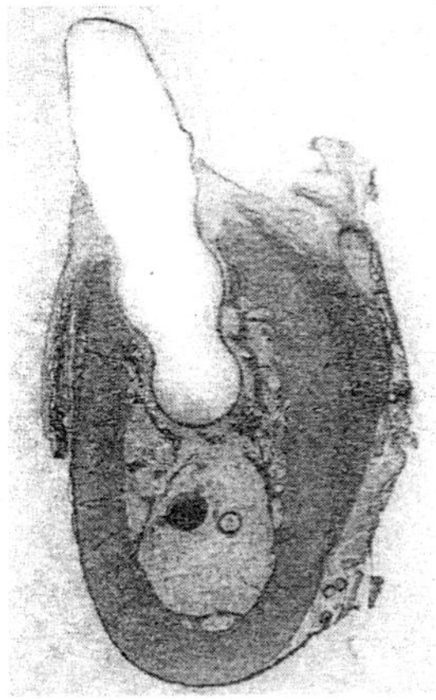

Fig. II-1. A specimen of an implanted artificial root (a dog)

This method is that by the energy based on masticatory movements with repetitious undulatory biomechanics of 0.2 mm width, the cementum (fibrous bone) and the periodontal membrane are produced in hybrid types, using the genes of transplanted cells from men or animals.

Three years after that, using porous ceramic artificial bone apparatus that combined the genes of the animal with artificially synthesized materials in muscles of mammals, and induced successfully hemopoietic nests and osteoblasts from muscle cells by the hydrodynamic stimulation of the humour firstly in the world. Even if you transplant this porous artificial bone into the subcutaneous tissue, nothing will happen. It was believed that it is absolutely impossible to produce hemopoetic nests or bone in muscles or subcutaneous tissues that lack bone, by artificial bone. By the development of these two artificial organs, I elucidated that the evolution of vertebrates had occurred by the response of living organisms to gravitational function, with mechanical stimuli centred. It was not until the landing drama (the second revolution) in the evolution of vertebrates that the work of hemopoiesis done originally at the intestine moved to bone marrow. The development of the teeth with gomphosis having the mammalian type periodontal membrane was brought about by the third revolution (birth of mammals). I discovered both vital tissues and structures developed at the epochs of evolution were produced by biomechanical stimuli.

Thinking Up Experimental Evolutionary Study

I transplanted an artificial bone marrow apparatus into a shark (the cartilaginous fish) typical of the primordial vertebrate with no bone marrow hemopoietic system, and induced hemopoietic nests, which was anticipation of what would come in evolution. (Fig. II-2.)

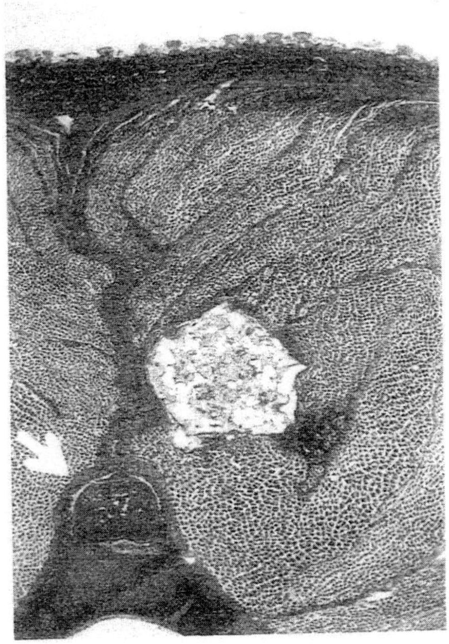

Fig. II-2. Induced haemopoietic nest after transplanting artificial bone marrow in a shark. (arrow: haemopoietic nest formed at the lower left of artificial bone marrow)

Once you understand evolution occurs by mechanical stimuli with a particular emphasis on gravity, you can cause the changes in shape or work of animal organs by giving them mechanical stimuli that would occur in evolution. This verifies that Lamarck's Use and Disuse Law is just right: Lamarck proposed in 1809 that a constant employment of the body organs determines their shapes and sizes. Therefore, I devised Experimental Evolutionary Study to cause evolution artificially by biomechanical stimuli. By letting a shark and a Mexican salamander (axolotl) land, and by causing the changes of their body artificially, I could verify that evolution occurred by mechanical stimuli with particular emphasis on gravity, including the change from water to air and complex stimuli such as rapid increase of oxygen in quantity. That is to say, when an animal lives long in response to the changes by gravitational function or in response to living media acting on the body, having the same inherited characters, the form of the cells in a tissue changes automatically in response to stimuli. This is called metaplasia. The great evolution of vertebrates is nothing but landing dram (terrestrialisation). The other transformations in shape proved to be caused by Wolff's law (the law of the transformation in shape according to function by repetitious loading in the skeletal system, also referred to as the law of functional adaptation in shape) by biomechanical stimuli alone according to the law of inertia with a lapse of time. In landing dram, the skin and the respiratory organ alone change by conversion from water to air. In the skeletal system, cartilaginous bone ossifies in response to changing from one-sixth G in the water to one G (gravity on earth). As a result of that, the shark responded by chance to the increase in gravity writhing in agony, its blood pressure rose, and also streaming potentials rose, which caused differentiation of cartilaginous cells into blood-forming cells and osteoblasts. This is also metaplasia. By this, the outline of evolution has been clarified. A phenomenon called metaplasia is a pathological term that cells of an organ change to cells of another organ by any stimuli. Since a human adult organism is composed of sixty trillion cells with the same inherited characters, it may be said that if the trigger of a specific gene is pulled, the cell can do metaplasia into anything. It was confirmed that the intrinsic factor in the law of evolution of vertebrates was a phenomenon called metaplasia.

Essence Of Life - The Mind And The Intestinal Visceral System

The essence of things is called the mind or spirit. What is the essence of life?

Erwin Schrodinger who made a start on molecular biology in the 20th century, thinking that the essence of life exists in hereditary phenomena, proposed biological research using physics as the method to publish 'What is life?'. Under his instigation, young physicists one after another began elucidating the mechanism of heredity using parasitic viruses on bacteria as the simplest heredity, and bacteria. Does the elucidation of the mechanism of heredity lead to solving the enigma of life phenomena? What has become of the life phenomena of non-hereditary one generation? A unicellular bacterium (prokaryote), a eucaryote (an animal composed of a grain of cell), and cultivated cells of the vertebrate are free of terrestrial gravity (1G), and live on even in ten thousand G or fifty thousand G, while the vertebrates composed of multi-cells can't live one day under five G. Is there any fundamental difference in the mechanism of life between unicellular organisms and the higher animals? Until this day in the 21st century, since the 20th, the question 'What is life?' has not been defined correctly.

While hereditary phenomena are near to the essence of life, there has to exist the essence of life, namely, the mind of life common to unicellular organisms and multi-cellular animals. The difference between living things and non-living things is nothing but whether to be able to replace old and stale parts of the body or not. Replacing is referred to as metabolism. This is a sign of living, and today it is called remodelling. The whole replacement of an individual is hereditary phenomenon. This replacement requires energy.

The new system of biology (molecular biology) based on the physical method by Schrodinger has lacked the viewpoints on energy metabolism in living organisms and energy action functioning from the outside on living organisms. So long as you can understand these, you will be easily able to define what life is. The constitution of living creatures is the mechanism in which parts of the individual develop, grow, and remodel, with energy vortex circulating, and which overcomes ageing. The whole replacement of the individual is hereditary phenomenon usually through the medium of reproduction. That energy vortex circulates, refers to cell respiration and glycolysis in the higher vertebrate, both of which utilize into life activities energy that can be obtained by aerobic or anaerobic resolution of nutrients absorbed from the intestine. This energy is obtained by digestion and absorption in the intestine as foods, and originally is produced with photosynthesis by sunlight in vegetables. Since oxygen is also absorbed from the branchial intestine in vertebrates, it is life that metabolizes with energy obtained by decomposing the substances absorbed from the intestine. The lung is also part of the intestine because it stems from the branchial intestine. Thus, all the energy resources presiding over remodelling the essence of life are dependent on the intestine. Since the other living organisms except vertebrates perform dermal respiration, the parietal system plays an important role as much as the intestine. In vertebrates, the intestinal visceral system is just in charge of the organs of digesting/absorbing foods and oxygen that are energy resources to support remodelling (metabolism and reproduction), the essence of life, and is the organ of the essence of life, or the resource of life.

Two Mechanisms Of Life

Higher living organisms have been born from the intestine, and sustained by it. Life does not exist without the intestine. Sure enough, the essence of life (the mind or spirit) exists in the intestine. The nerves distributed at the intestinal epithelia transmit the situation of the inner lumen and visceral muscle of the intestine to the muscle of the body skin (parietal muscle). By accepting the information, the parietal locomotive muscle moves. Namely, after the intestine's recognition the body locomotive muscle moves toward its wishful direction. This is the beginning of an impression when the animal moves by sensing. A solidly built parietal body composed of nerves, muscles and bone is just a vehicle to carry the mind in the intestine. The essence of life is the renewal of cells occurring together with the rotation of energy's vortex, namely remodelling, which is the reproductive system at the end. Thus, the essence of life turns out to be a mechanism to live against aging. The mechanism of life has two important aspects as previously mentioned. One is the energy's vortex, which is the side of fuzzy material energy having no mass. This is called 'Kun' (vanity or emptiness) in Mahayana Buddhism. It refers to body temperature, bioelectricity, bioluminescence, and life energy that the animal has as the source of motion, which conventionally has been classified as the sciences of function such as physiology, biochemistry, or pharmacology. The mind, spirit, thought or soul is in fact a kind of energy without mass, but until now the mind, or thought, has not been recognized to be life energy. The other is renewal, which refers to the development of cells, tissues and organs, growth and remodelling. Namely, it is the side of the body parts in animals, which is composed of substances with mass. It is called 'Shiki' (matter' in Mahayana Buddhism, of which the science is morphology. In the view of the world in Christianity, the mind, spirit, thought or spiritual phenomenon, having the same energy as body temperature, has been confused with the body with mass.

Morphology And Science Of Energy

It was the famous poet, Goethe that established morphology in modern science in 1795. He defined morphology as the 'nomination of the shape and organ of organisms, and elucidation of the law of transformation in life morphology' namely metamorphosis. He was the first scholar to have recognized that 'there is no aim and no purpose in living organisms' in the 19th century Christian world through his life-long research into morphology. To be sure, there is no purpose in life, but the mechanism of life itself is that of self-remodelling. It is life that feels unequalled pleasure in remodelling. This mechanism is life, and the essence of life exists in the intestine. Therefore, the regeneration of cells and reproduction of an individual by remodelling may be considered to be the purpose of life.

Life Phenomena and Electrical Phenomena

Life phenomena are transfer of ions in a water solution, namely the mechanism of electrical phenomena. Since it is the finest electrical response system in the universe, naturally life has no purpose or aim. Energy metabolism conjugated with remodelling, namely, circulation of energy vortex together with metabolism in water- soluble colloid is life. Metabolism naturally generates heat. This is temperature. Remodelling of an individual whole is reproduction. Renewals begin by fertilization of a sperm and ovum, which are haploid prokaryotes manufactured by miosis of one cell. Living things are electrical phenomena of water-soluble colloid of substance with mass, accompanied by mass-less energy equipped with it. The vortex of life energy seeks by itself for remodelling. The mechanism of taking in energy of remodelling and nutrients with mass lies in the intestine, and the ultimate mechanism of remodelling is reproduction. Reproduction is also undergone by the intestine-abdomen in mammals.

The purpose of Goethe's morphology is to elucidate the law of evolution. Goethe said 'If you want to know the essence of an organ, all you have to do is to trace back the origin'. It was 14 years earlier when Lamarck of France proposed the Use and the Disuse Theory as an evolutionary theory in 1809. The essence of an organ can be known by its function. Therefore, if you want to know the essence of the function of an organ, all you have to do is to trace back its origin. Functional abnormality is the so-called intractable disease (immune disease). Therefore, the essence of immune disease of an organ can be known by racing back its origin.

Three Enigmas Of The Vertebrate

Though the great masters in the world of biology, Cuvier, Linne, Lamarck, and Goethe, had studied the morphology of living matters including vegetation, what had captivated most their minds was the vertebrate, the distinguished family in the animal world.

Vertebrates have three riddles to be solved. Why does evolution occur? How is the mechanism of the immune system? Why does the mechanism of haemopoiesis in the internal organ move to the parietal system in the higher animal alone? The intestine that digests and absorbs foods is originally the mechanism to produce blood that becomes flesh afterwards. The author has clarified the enigma of the development of bone marrow haemopoiesis using an artificial bone marrow haemopoietic apparatus. As the result, when the sharks of the cartilaginous fish were exposed to the earth environment to be obliged to land, their cells constituting their body parts turned out to have underwent automatically metaplasia, facing gravity and the changes in life media. So far, three different lateral aspects have been observed in the same phenomenon of evolution, and these three enigmas turned out to be derived from gravity-response. In short, with the same hereditary characters, cell forms changed by the changes of functioning stimuli by metaplasia. The cause to induce these three enigmas turned out to be weightless energy, which is biomechanical stimulus in a broad sense, namely, the function by physical (energy) and chemical (substance with mass) stimulus. By this function, the trigger of cellular genes is pulled. While the development of bone marrow haemopoiesis was clarified to have been caused by function of gravity energy and evolution was elucidated, it is not easy to understand the mechanism of the immune system. The reason is that we can know only the fact that the occurrence of major histocompatibility antigen to decide whether transformed grafts are adhered alive or refused, depends on gravity function at landing drama that coincides with the occurrence of bone marrow haemopoiesis.

The mechanism of the immune system, or the mechanism to conquer diseases, does not lie in only this major histocompatibility antigen. Among three enigmas, namely, among three functions, the most troublesome is the mechanism of the immune system. The system to conquer diseases seems to lie in the mechanism of remodelling occurring with rotation of energy vortex. Accordingly, before we clarify the whereabouts of the mind and spirit, we have to clarify the mechanism of the immune system. The mechanism of the immune system can be solved by tracing back the occurrence of histoimmunity and the functional modulation of organs.

Oral Brain, Anal Brain and Intestinal Brain

An old proverb says that even a worm will turn. In higher animals, first of all, the intestine occurs and gradually complex constitution develops. It looks as though the mind or soul dwells in the animals having the intestine. The elucidation of the evolution of vertebrates may lead to 'embryology' of the mind, soul, or spirit.

To think primitive animals schematically, firstly coelenterata were born from the intestine, and with coming of the mouth, Deuterostomia followed them. This is the beginning of the vertebrate, which consists of the double tubes of inner side and outside. The outside tube is composed of ectodermal epithelium and muscles moving body (mesodermal parietal muscles), and the inner side tube is composed of endodermal epithelium (intestine) and visceral smooth muscles (mesodermal visceral muscles). The outside and inner side epithelia are lined with muscles respectively.

The epithelium is connected to the cell membrane, and lets the respective muscle inform the surrounding status of the epithelium. This is the nerve, which is governed by membrane potential (the electric potential which exists on the two sides of a membrane. minus during stationary period, plus during excitatory period) and ion channel (work to pass specific ions into a cell). The epithelium is lined with a layer of cells with the cell membrane continued, and has no blood vessels because of not self-moving apparatus. It is the epithelium that moves accompanied by muscular movements. Both muscles attached to the exoderm and muscles attached to the endoderm abound in blood vessels. The mesodermal system that produces the original moving apparatus characteristic of animals, consists of the vascular circulatory system and muscle system. Visceral muscles have the cartilage of visceral skeleton, and the somatic system has bone, cartilage, and tendon which are the system supporting muscles. This mesoderm is governed by hydrodynamic streaming potentials.

It is hormone governed by the nervous system that acts as mediator between the skin-brain, and nerve tissue in the ectoderm, and the somatic exercise system in the mesoderm. It is the cerebrospinal central nervous system that is the bundle of the nerves that combines the skin nerve with the intestinal nerve. Originally, to look at each segment in the spinal cord, one segment has one brain. An ascidian, a monosegmental protochordate, becomes conceivably a multi-segmental animal by gene duplication (gene unit becoming diploid or triploid) and when each segmental connects in sequence and becomes the spinal cord, the central nervous system is divided into three kinds. At the exit and entrance of the intestine, the nerves of the skin and the intestinal duct combine. Accordingly, there exist the oral brain=branchial brain (the author's coinage meaning the brain in charge of respiration, mastication, digestion-absorption and sense-exercise), and the anal brain=excretory brain (the brain in charge of reproduction, urination, and excretion). Halfway between them, there exists the intestinal brain that is pitch dark.

Since life is born from the intestine, the body's status is always determined by the conditions of the inner cavity of the intestine. It is the hallmark of the animals that they move with employment of body muscles according to the demand of the intestine. The ways of muscular movements of the intestine, or the desires of the intestine appear somehow or other to be the mind or soul.

Development of the Mind and Spirit

In order to clarify 'the development of the mind and spirit' first of all, we have to elucidate the evolution of the internal organs or other organs concerned. It is necessary to clarify the evolution of the intestine, brain, nervous and cardio-intestinal system, as mentioned so far.

Since I clarified that evolution has really occurred by animals' response to gravity, I have researched into phylogeny concerning evolution and comparative morphology using from an ascidian to a hagfish of Cyclostomata, shark, axolotl, xenopus, newt, lizard, quail, rat, mouse, dog, and monkey. As a result, I have found out a shark to become a mammal, and also discovered that the tongue was derived from respiratory visceral muscles, the diaphragm had existed in a shark already that developed from the shark's bottom of the pericardial cavity, and clavicle was the elongation of the branchial arch bone. I will describe them in detail later.

Only these morphological researches cannot clarify the development of the mind or spirit. 'The face and mind' or 'the body and spirit' might be able to be solved barely. However, since the mind or spirit is the same life energy as body temperature or perspiration, it has to be approached from the clarification of function, namely, work of cells.

Clarification of the Mechanism of Immunity by Transplant Experiment

Among functions, the most difficult is the mechanism of immunity. Today, it is immunology by self – not self that has fallen into great confusion and disorder. The immune diseases that were curable in the past, have not been cured by it. Immunology by self and nonself must be wrong with something essential. Therefore, I decided to elucidate from the point of evolution the development of "the transplant (histogenous) immune system" that is the main theme of self and nonself immunology. While Haeckel's Recapitulation Theory 'ontogeny repeats phylogeny' is a theory of morphology, the development of the function-centred histogeneous immune system also appears to be involved with gravity like that of the bone marrow haemopoietic system, because major histocompatibility antigen exists mainly in the membrane of the leukocyte, and the development and differentiation of the leukocyte are involved in the evolution of the haemopoietic system.

The organs or tissues of mammalian foetus including human embryo can be transplanted into adult organisms. At the time of an embryo or foetus, the genes of major histocompatibility antigen are at rest without functioning. This is referred to as immunologic tolerance. In other words, antigens that are proteins having gene-function, can't be produced. Genes are the information of codes to produce protein.

When I elucidated by the research into the development of the haemopoietic system that haemopoietic tissues moved from the spleen to bone marrow cavity by gravitational function, I got a confident idea that the trigger of the genes of major histocompatibility antigen must have been also pulled at the same time in response to gravitational function. To verify this can be viable by quite simple experiments. The author succeeded admirably in skin graft of the same kind between sharks, between a shark and a cyclostome, between a shark and Xenopus, and between a shark and a mammal, which was considered conventionally to be impossible. I was much pleased to find out that even shark's cornea, brain, and intestine were successfully transplanted to the respective organs of mammals.

Establishing Clinical Phylogemy

I clarified the following fact: major histocompatibility antigen that is the pivot of self and nonself immunology, is not the system to distinguish stale, cancerous, or benign cells by compatibility antigen existing in the membrane of the leukocyte. Namely, the mechanism in charge of metabolism in the membrane of the leukocyte is major histocompatibility antigen. So, I devised a practical medical method called practical phylogenetics in which the development of diseases was correlated with the interrelationship between organs obtained by the research into evolution, structural defects in the body, or biased or erroneous usage of the body, by observing patients with various kinds of immune diseases. I integrated Gravity-corresponding Evolutionary Study with medicine. In order to avoid the error of self and nonself immunology, the author clarified in detail the differences between healthy persons and patients with immune diseases from the standpoints of energy metabolism and energy function by observing diseases. Namely, I followed medical procedure to cure diseases in the same way as in the birth of bacteriology. As a result, immune diseases turned out to be caused by the modulation of energy metabolism due to extensive and various kinds of factors.

Energy-metabolism and Immune Diseases

Since energy metabolism in man is mainly dependent on respiration, modulation of the function of mitochondria being a small organelle, the existence of parasites (bacteria or viruses) to impair directly materials to produce mitochondria, lack of Vitamin B, C, and coenzymes essential for the metabolism of energy small organelle, lack of oxygen, and so on, turned out to develop immune diseases.

In practical therapeutic examples, a senile man having undergone internal medicine under the diagnosis of leukaemia with severe dementia and rash, recovered completely by extracting ten teeth with severe periodontitis, and a female of sixty-seven years old to lose consciousness often with severe urticaria, recovered by changing oral breathing to nasal breathing. Infection of bacteria causing periodontal disease into leukocytes of the whole body develops rash and dementia. Such diseases mistakenly diagnosed as leukaemia can not be cured. The jawbone surrounding a tooth is derived from the skin called the dermal bone at the age of the shark. The infection of the haemopoietic nest of the bone by bacteria causing periodontal disease developed the infection of leukocytes in the body. The brain, eyes, and nerves are derived from the ectoderm like the skin. Namely, all the organs having the same origin are infected by leukocytes carrying saprophytic bacteria. Saprophytic bacteria have caused rash and erosion in the mesenchymal glia of the brain. That old lady's fainting away was caused by rash and erosion in the mesenchymal glia of the brain, with exhaust gas from the car having been absorbed from the tonsilla and lung by oral respiration.

A male of forty- three years old in a lapse of 13 years after having lost his eyesight, visited me for treatment, after having heard the voiced version of my book 'Health is decided by respiration'. Since he suffered from atopic dermatitis and had the face of oral-breathing, I asked him that he had played excessive sports before onset. He answered, 'I had played American foot ball and rugby football, and three years after stopping the sports I noticed loss of my eyesight because during driving traffic signals had disappeared. This is the phenomenon that atopy occurred at the retina by excessive exercises. Aerobic saprophytic bacteria enter from M-cell's (Microvillicells: mechanism of the intestinal tonsilla to produce secretor Immunoglobulin A) follicle, cause atopic dermatitis. Atopic dermatitis can be cured simply by rest and nasal respiration: in addition to improving them, warming the abdomen and eating by chewing well to take good care of the abdomen let his eyesight recover to the extent he distinguish slightly between colours. Thus, immune diseases in the skin-centred parietal system destroy nerves in the brain, spirit or mind.

The Mind and Mental Diseases

Thus, ignorance of the traits and functions of the intestine leads to impairment of way of thinking, thought, and spirit by influencing the somatic system such as the brain or skin directly. On the other side, oxygen deficiency in the portal vein of the intestine due to lessening function of the intestine on various reasons, inactivate the internal organs, which often causes illness in one's heart. Visceroptosis due to lack of "bone-rest", cold food-mania to cool the intestine, rough way of eating without enough biting, and oral breathing, impair visceral muscles in the intestine to cause menstrual pain, endometriosis, headache, migraine, neurosis, and depression, by modulation in the parasympathetic nerve.

On the other hand, short time sleep in which muscles in the parietal system extremely strain or are exhausted, oral breathing, chest respiration, poisoning by cold food to cool the intestine, irregular meals, and symathiocotonia, cause panic syndrome, hyperpnea, and schizophrenia.

Illness of heart and spirit due to modulation in the autonomic nervous system is also immune disease caused by disturbance in energy metabolism.

Those suffering from these immune diseases lack bone-rest, cool their intestine by cold food mania, impair their intestine, and what's more, respire by the mouth without fail. It is only human beings after at the age of one among mammals that can respire by the mouth. It is the weakest point in the human body structure. Oral respiration and cooling the intestine make ceaselessly resident flora from M-cells in the tonsilla by the leukocytes circulate through the body. Lack in bone-rest worsens the function to produce blood in bone, and leukocytes are worn out. As a result, the resident florae start playing all sorts of pranks here and there in the body.

The theory that major histocompatibility antigen discriminates self from nonself is 0immunologists' great misunderstanding. It is the mechanism for lymphocytes to destroy old cells and tumour cells by discriminating with tired or broken cell membrane at the time of remodelling in which a billion cells remodel in a day and which was called often metabolism. The author discovered that illness has occurred by mass less energy, and immune diseases have occurred by nonpathogenic saprophytic bacteria or proteins due to paying no attention to the characters of the intestine. Elucidation of illness by this method leads to treatments for heart or mental diseases.

Both heart and spirit are a kind of life energy. Life energy refers to superheating, electric current, electromagnetic wave, and light that are generated by the activity of cells of life machine, as an organism consisting of protein, nucleic acid, glucose, and lipid. By the energy, in multicellular animals, leukocytes digest, detoxificate, anabolise, catabolise, toxic, non-toxic, or helpful substances such as nutrients, oxygen, toxin, micro-organisms, parasites and so on that always enter into the body from the body surface and the intestinal epithelia.

Thus, energy metabolism rotates, which serves to remodel. This power is the so-called immune power. Without digestion, you are susceptible to infection or toxin, which is the infectious immune disease. All these facts are entirely beyond the method of life science based on conventional "the law of conservation of mass". Contemporary medicine trying to cure by mass-substance alone based on the understanding of universe in 19th century is out of date and quite unhelpful. The time will surely come soon that a superior doctor can prevent and cure illness except surgical diseases by elucidation of evolution. Thus, by introducing the principle of the conservation of energy into life science, I have integrated evolutionary study into the immune system, and established a new immunology. Now, we are just in time for getting the way how they develop concerning with heart, mind, soul, spirit, thought and gigong that are considered to be the most mysterious in human beings.

Mind-existing Cells and Organs

Once mind is recognized as life energy, the following questions are raised: What part of the multi-cellular animal does mind exist in? Does mind exist even in the mono-cellular animal? Mind, to be natural, exists in life activity in life machine, namely, in functions consisting of nucleic acid, protein, lipid, saccharide, salt, phosphoric acid of the mono-cellular animal. In the multi-cellular animals, mind exists in the function of the organs of the intestinal visceral system being source of life, namely, peristalsis of the intestine in a broad meaning and leukocytes that have the function of digestion and absorption and retain the figure of the protozoan consisting of the mono-cell more than 500 million years. Since mind exists in motor function in the smoatic motor muscle system of higher animals in which the sympathetic nerve and pyramidal tract system have developed, life energy called mind does not develop in the animals from primitive vertebrates to lower protozoan which do not have the sympathetic nerve and pyramidal tract system.

Thinking of the immune system from the view point that mind and spirit turned out to be life energy generated by the highest function of the cell and cell group, energy metabolism of higher life next to mind and spirit, or the function of leukocyte-centred blood turns to be immune power. I have clarified that in this immune system, disturbance in remodelling occurring with rotation of vortex of mass less substance and infection by resident flora that should be non-toxic, develop immune diseases. By this, I have devised the method to cure radically intractable diseases of which treatment was unknown from the aspect of energy.

Chapter Three

Embryology of Mind and Spirit
The Respiratory Organs, Intestine, Nerve and Muscle
Preface

Elucidation of evolution of life might clarify whereabouts of mind, and where it is born from.

However, if evolution of vertebrates can not be elucidated perfectly, it would be impossible to clarify them. The author, as previously mentioned, clarified the enigma of the sudden development of the mechanism of bone marrow haemopoiesis in the landing drama in the second revolution of vertebrates, and the enigma of the development of the mechanism of teeth with gomphosis in the third revolution of vertebrates. Using ceramic artificial bone and artificial dental root, I developed the method to induce the cementum, fibrous tissue, haemopoietic tissue, and bone tissue on the surface of ceramics by hydrodynamic stimuli of the body fluid being biomechanical energy, namely, by borrowing cell genes of transplanted recipients.

The change in the employment of the animal body causes the change of energy acceptance with having the same genes to change the shape and function of the animal organ. Given that only this behavioural pattern can be transmitted by any means, the transformed shape will be transmitted to the next generation with the same genes. Namely, the trigger of the same genes is pulled by the change of energy, and this transformation brings about the cells of other shapes or functions. As previously mentioned, it is called metaplasia by the word of pathology. If behavioural patterns are transmitted by any means, for instance, by education, behavioural patterns of children change by memory. As a result, according to behavioural patterns the forms of the hand, leg, jaw, and the face change gradually along with growth. Thousands and tens of thousands of the repetition bring about running after the change of shape by copy-error occurring once in a million times in the reproductive cells, namely, by mutation, and the unintentional change of genes occurs. This is molecular evolution.

There appears slight change in protein by this. These accumulation does not let the behavioural pattern return to the one of the old species, even if the behavioural pattern is returned to the original one at the initial stage of development. Thus, when a new behavioural pattern is established, shapes change and a new species segregates with the same genes. After the behavioural pattern is transmitted for long period and over generation, genes change little by little with the lapse of time, and finally a new species having the different shape and gene segregates.

The reason why everyone succeeds in letting axolotl land, is that without water the trigger of the cell genes changing into the reptile type is pulled. However, if axolotl continues to live only in the water more one thousand years, it will not return to the reptile, even if with reduced water. It is owing to molecular evolution by running after. Thus, acquired characters are to be inherited. It is only the shape of the external nose that differentiates the mammals from the amphibia, reptile and bird among the part of the face. The mammals have the nasal shape of a hieroglyph, and the others have two small nostrils. This means that the ancestors of the mammal and the other species are different.

As the birth of Dolly, a clone, shows that basic mesenchymal cells constituting our body have enough genes to produce every organ of our body. The trigger of genes is pulled by electric current, electromagnetic wave, or super sound wave in a case, and by oxygen or chemical substances in some cases. Evolution has occurred without fail by the change of a fixed behavioural pattern with retaining the same gene characters, on the basis of three kinds of collaboration such as streaming potential developing at the skeletal system, membrane potential developing at the nervous system, and its secretion, hormones. If such is the situation and you really want to understand evolution, you have got to realize well by all means what energy is.

The basic principle of natural science is "elucidation of the law hiding in the background of intricate phenomena". This was proposed by Goethe who originated morphology, and by Lamarck who elucidated the law of evolution at the beginning of 19th century. There is a hierarchy of the technique for science called strict "study, art, logic, and law" in learning. Modern life science raises the following five problems:

1. the constituent law of the universe and its relation with life, 2. subdivision of science and reductionism, 3. materialism, 4. pitfall in molecular biology, 5. absence of vertebrate zoology. The constituting elements of the universe are the following five ones: 1. time, 2. space, 3. materials with mass, 4. gravitational and mechanical energy, 5. wave and thermodynamic energy such as electromagnetic wave. However, even in today, not only life but also every phenomenon is explained only by materials with mass. That is why many people could not predict that the pacemaker or instruments of airplanes become out of order by the cellular phone. Energies such as gravity, electromagnetic wave, or thermodynamics and also the function of time are essentially important for the metabolism of living things. At length we could introduce the achievements of 20th century's science into the science of evolution at the 21st century of today in 100 years late. Let us consider the phylogeny of mind and spirit from here through the evolution of the respiratory organs, intestine, nerve, muscle and the like.

Evolution Of Life Viewed From The Nose

The Origin Of The Face

The human face is the complex organ representing life that represents personality, character, spirit and mind. As the word called a cast of countenance shows, the structure of the face itself shows one's soul, namely, one's mind. The external nose is centred on the face. Vertebrates have three kinds of brain. Since the brain is the bundle of neural tubes that combines the nerve of endodermal epithelium in the intestine and the one of ectodermal epithelium in the skin, one is the branchial brain on the oral side, and another is the excretory brain on the anal side, and the other is the intestinal brain engaging in digestion and absorption at the in-between pitch-dark intestine. The face, urinary organs, reproductive organs, and anus are the organs in which intestinal muscles protrude right under the skin of ectoderm. Since mind or soul exists in the intestinal visceral system, the shape of the face and reproductive organs represent mind and soul. That is the reason why a man is in rut and his pitch-dark intestine and excretory intestine operate if he looks at the shape of the face and reproductive organs in charge of the oral and anal brain.

Goethe, who originated morphology, had also been plagued with the wonder of emotion and feeling evoked by the human figure and feature through the study of cranium morphology through all his life. Packard, Cope, Osborn, Simpson, Marsh et al in America studied the evolution of mamals using cranium and teeth from the end of 19th century to the beginning of the 20th century, and advocated Lamarck's theory. These scholars also in America called neo-Lamarckist could not change evolution theory to evolutionary study that elucidated by the law of causation based on gravitational function in the 20th century.

The face is the uppermost segment, which origin is an ascidian (protochordate) consisting of mono-segment. (Refer to Fig.V-1). The face of an ascidian has every organ, and it has the branchial intestine, the gastrointestinal tract and the excretory intestine.

This mono-segmental ascidian repeated gene duplication several times, and an multi-segmental individual composed of chain-salpa type's segments tied in a row did cephalad movement. As a result, according to the Use and Disuse Theory, the vertebrate had the mouth similar to the mono-segmental ascidian, the branchial intestine, the intestine in charge of absorption and digestion, the excretory intestine, the anus and the tail. (Refer to Fig. III-2).

Work Of Mitochondria In Cell Respiration

Protozoa far long before an ascidian such as Malarian pleomodium, paramecium, and amoeba, live by the mono-cell. They have , according to species, radiant corresponding to the eye, or the mouth, pseudopodium, chemical substance sensor similar to the sense of smell or taste, and cilium corresponding to the sense of touch. Namely, they have the anlage of the face and every organ of the body by a grain, and naturally mind and memory. Protozoa also employ a part of genes to the formation of various cell organelle. However, since mitochondria are prokaryote parasitic to eucaryote, their nuclear genes produce only polymerase duplicating DNA and RNA of them. Multi-cellular animals produce one multi-cellular organ by a part of genes. In multi-cellular vertebrate, mitochondria are in charge of cell respiration being the centre of energy metabolism.

The vertebrates are chordates having osseous spinal cord with various ossification, and their characteristic organ performs intestinal respiration. The elucidation of the relationship among bone, branchial respiration, and cell respiration, leads to the enigma of the vertebrate. Since the face has bone, teeth, nose being the entrance of branchial respiration and is composed of branchial intestinal muscles, by studying of the face alone, evolution, immunity, and the enigma of bone marrow haemopoiesis can be solved.

Use and Disuse, and Wolff's Law

The shape and structure of the vertebrate is determined according to Wolff's law when its way of movement is fixed. This is because movement of a part of the body induces muscle accompanied by nerve, and the cartilage or bone of the skeleton, has the characteristic that its shape changes according to Wolff's law. Wolff's law is an empirical law: when the function of hydromechanics flowing within the body changes by movement of the body, the change of streaming potential is caused, by which potential the trigger of the genes of muscle cells or endothelium cells of the blood vessel are pulled, and various cytokines are induced, by which the shape of the muscle-skeletal system changes. In this occasion, blood vessels are induced together with proliferation of muscles. Wolff's law can be considered to be a part-law of Lamarck's Use and Disuse to which only skeletal muscles and the shape of bone are adapted.

Thus, the vertebrate changes its shape in accordance with motion with the same heredity. When food and life medium from water to air change under extraordinary disasters, the shape of the tooth, face, and hand and foot of a certain animal changes. Among behavioural pattern, eating plays the most powerful function in life. This change of shape is reproduced in ontogeny. Tracing back the ontogeny of vertebrates, both of the shape of the head and face change from primordial mammalian type to horned shark one, and moreover, the branchial apparatus becomes ancient lamprey type, and at the first stage the head and face becomes ascidian type consisting of mono-segment. This means that the origin of the vertebrate is really a shark (Cyclostomata), and an ascidian.

The Origin Of The Vertebrate And The Nose's Shape

Tracing back the shape of a mammalian foetus results in the primordial cartilaginous fish-type. In birds, reptiles, and Amphibia among vertebrates, their viviparous shapes apparently differ slightly from the ones of the mammals.

Let's explore a primordial cartilaginous fish similar to the shape of 32nd to 34th day of a human embryo. You'll easily find a dog shark of which scientific name is Heterodontus Japonicus. The mammal is the animal having nursing, sucking and masticating system, as it grows up. The teeth performing mastication are Heterodontia: the teeth with their shapes differing according to the site of the jaw. Accordingly, the scientific name of a dog shark means a shark having mammalian teeth.

This shark differs from a common shark in all aspects; this shark is destined to become really a mammal.

First of all, a dog shark masticates crunch-crunch shrimps or turban shells shell-whole. It has already three kinds of teeth. Both the upper jaw and the mandible are closely equipped with shark-type incisors, shark-type canines, and tile-shaped molars. Its endoskeleton has partly ossified and bone marrow haemopoiesis has developed. Its heart is large, and there are cartilaginous bones in the dorsal side of the pericardial cavity, and since the oral side is open, the thymus and lungs go backward and enter into the pericardial cavity. Its nostrils are open to the oral cavity. The shape of its external nose is the one of the mammalian nose.

Namely, its nasal shape is already a hieroglyph, Ω. In the buccal region, masticatory muscles derived from branchial muscles similar to masseter muscle. This shark is eaten as a higher cuisine at Bonn-festival and New-Year in Kyushu and Hiroshima districts; it is very delicious without unpleasant order of the shark just like sturgeon meat.

The author succeeded in transplanting shark's tissues or organs into mammals with every operation successful, and verified Biogenic law by Haeckel that ontogeny repeats phylogeny, from the view point of the histo-immune system, whose theory is originally of morphology.

Accordingly, though this theory would have been verified even with comparison by shapes, the world had not really believed in Haeckel. That's why it had overlooked the fact that the face of Heterodontus Japonicus had the same face as that of the embryo of the mammal including the man. (Refer to Fig. III-1.)

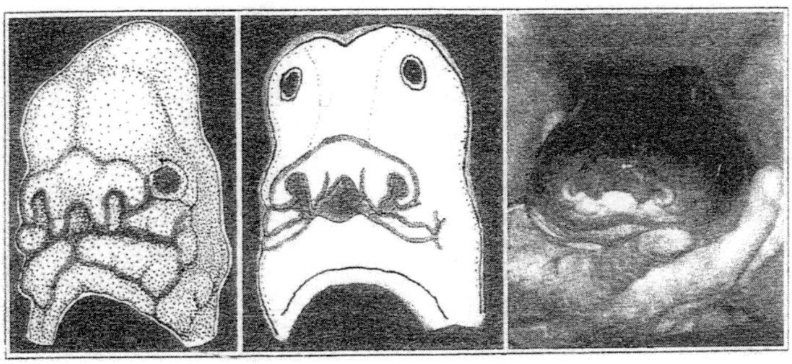

Fig. III-1. The face of Heterodontus and human embryo (Left: a human embryo, Right: Heterodontus, Photo: front face of Heterodontus).

The external nasal shape is of mammalian type, and separate nostrils put together at the midpoint by employing strongly jaws for mastication: extension in the middle portion of the nose closes completely and becomes a sinew of a dog or cat. Heterodontus Japonicus's nostrils open in the mouth and perform branchial respiration by the nose, while the other sharks7 nostrils open in the skin only with the function of the sense of smell. The other sharks perform branchial respiration by the mouth. In the reptile, birds, and amphobius, the bronchus opens in the tongue, and the nose opens in the mouth just like the shape of Heterodontus Japonicus's nose, and they perform oral-respiration. In a hagfish, the duct leading from the nostril to the throat is already open and performs branchial respiration through the nose. Its body surface is covered with slimy mucus. Among sharks, only Heterodontus Japonicus's body surface is slimy mucus. After all, the prototype of the human face can be searched for Heterodontus Japoponicus. The external nasal shape is corresponding with that of 34th day's human embryo in detailed parts; by and by, embryo-period (the period in which the shape of mammalian face completes by reproducing phylogeny stage) finishes and at the stage of foetus (the embryo from having the human face to birth), every part of the mouth unites to complete the human face. At the 38th day, the external nose and the groove existing in the surrounding of the mouth unite to complete the human face. There is no living animal today with not united at the 36th day. It is in a rabbit that concrescence is insufficient and a sinew remains. It remains as a sinew in a dog, cat or lion. It is the human cleft lip, cleft palate, oblique facial cleft, and transverse facial cleft that a baby is born with cleft condition. The cause for these development is that just at the time of concrescence from 36th to 38th day after fertilization, the mother womb suffers from insufficient oxygen, and the process of development is disturbed; concrescence does not occur and reversion occurs. The external nose of Heterodontas Japonicus hardly moves. However, while it makes a landing and continue to masticate by wriggling, branchial intestinal respiratory muscles become linked with external respiration during mastication. The mammals perform diaphragm respiration by expanding nasal wings, while many Japanese in the habit of oral respiration are weak in the above-mentioned respiration and die by performing nasal wing respiration at their death. After a lapse of 500 million year's evolution of vertebrates, no matter how strong you let respiratory muscles of the nasal wing move, oxygen-respiration can not be performed at all in the muscles far from the external respiratory branchia, but only muscles move. This is the reminder of branchial respiration in the Pleozoic era that is reproduced but a short while at one's death.

The human beings that perform external respiration by the lungs through the nose having the external nasal form of the mammal seen in Heterodontas Japonicus, intake oxygen to blood corpuscles by inhaling and exhaling, and emit carbon dioxide. Mitochondria being the parasite of prokaryote use this oxygen in cell respiration to have charge of energy metabolism. The mammal conquers aging with development, differentiation, growth, and metabolism by using this energy. We live by depending entirely on the ancient parasite, mitochondria.

External Respiration: The Nose And Lung

It is the external nostrils and nasal cavity that feels firsthand the regular rhythmic exercise of expiration and inspiration, and thoraces feel it from the viewpoint of the visceral body. The branchial intestine is the one expressing mind. Your heart leaps, or your heart hurts with choking by the respiratory intestine. In the mammal, external respiration to support life can be obtained alternatively either by 'life respiration' by the parasympathetic nerve through the nose with the diaphragm, tongue, and heart derived from the branchial intestinal muscles, namely, unconscious 'involuntary' respiration or 'voluntary' respiration, namely, respiration with will by the thoracic movement by extrapyramidal muscles innervated by the sympathetic nerve. (By Shigeo, Miki)

Among the mammals, there are some evolved animals that do not quite retain the prototype of the nose. They are a part of whales that were reduced into living in the water, and elephants. It has been proved by the shape of an embryo of the elephant that the elephant had been living in the water. What is the reason why the external nose transforms from its prototype? It is because of the way to exercise the body mentioned previously. Ordinarily, air enters naturally from the nostril by involuntary respiration as long as you live in the air, while in the water you can not inhale air unless you let your nostrils out of the water every time. Since a sea lion and a fur seal take their nose out of the water together with their head, their shape of the nostril and form of the external nose do not change no matter how long they live. However, since whales and dolphins swim with their back and belly wobbled, their nostril respiration and their body movement differ quite from those of fur seals. Most bones of the leg and pelvis are disused and disappear by the propelling power of dolphin-typed movement. When an animal inhales with its nasal head out of the water and continue to swim dolphin-type, its nostril moves caudally 0.0001 micron per day by the law of inertia, and after the lapse of about ten million years, its nostril moves thirty centimetres toward the top of the head. When dolphin-type movement is extremely slow as in manatee, its nostril does not move and become oblong, which is well corresponding with mechanical response. Evolutionary science corresponding to gravity can calculate backwards, by the changed form or function, functioning period, strength of dynamics such as gravity, and degree of changed matter property or chemical characteristics. Thus, the fins and the nasal shapes of whales differing from those of other animals in the swimming style, transform over time by retaining the same gene characters with those of other animals. But, the site of the nostril stops at the perpendicular portion emerging out of the water over the head. It is because they do not let their head move backward over its site when inhaling.

The Evolution of Life Viewed from Breath
Internal and External Respiration

Breath is referred to as respiration, namely, energy metabolism. External respiration refers to respiration as far as lungs, and internal respiration refers to cell respiration which is rotatory work of oxidative phosphorylation TCA cycle performed by mitochondria being an organelle. It is diet and sleep that directly influence cell respiration. By diet, mitochondria are supplied with energy resources, namely, macromolecular substances such as sugar, lipid, and amino acids that are absorbed in the intestine. Slumber is not to sleep, but to give bones a rest in order to remove the load on bones in the standing or sitting position, that is, to release potential energy by gravitational function, and to maintain blood pressure at 90 millimetres of mercury that is of standard in the vertebrate. Collagen, cartilage and bone consisting of skeletal materials are made of materials essential for the metabolism of mitochondria, and since they are metabolites of mitochondria, in the case of largely deviated gravity function in the skeleton, energy metabolism is impaired.

Namely, respiration, diet, and bone-rest have been fundamental elements of life that directly influence energy metabolism of the vertebrate. Even in the case of lack of bone-rest, the function of mitochondria in charge of cell respiration is impaired. Here, again, I describe the definition of the vertebrate. The vertebrate is a chordate having osseous notochord in spite of the degree of ossification, and an animal performing intestinal respiration. A common animal performs cutaneous respiration as a rule. The vertebrate also often performs cutaneous or mucous respiration. The amphibian's external branchia is skin, and it respirates even with the abdominal skin. Since this is a phenomena in which erythrocyte-haemopoiesis suddenly occurs in the epidermis of the skin and subcutaneous mesenchymal tissue, it is derived from gene expression of mesenchymal cells by apparently stimulating oxygen.

A soft-shelled turtle hibernates in the water during water, and its mouth is filled with cilia in which its tongue membrane did metaplasia branchially. Hibernating mammals have brown lipid. Lipid is fully filled with mitochondria, and becomes brown. When blood corpuscles take into oxygen at the lung or skin by external respiration, mitochondria pass it to cells and consume oxygen by internal respiration.

The Essence Of Immunity And Energy Metabolism

Let's consider 'what is life?' and 'what is immunity?', thinking of 'breath' through evolution that is the essence life. Viewed from the law of the conservation of energy, life is the complete open system in the universe. The reason is because life is directly linked with the energy resource with mass less energy pouring just into living things. If only life exists in the solar system, solar energy being mass less substance enters into or exits from living things freely. Judging from the law of conservation of mass (constitutive law of the universe in the 19th century), life is in the semi-closed system. However, misunderstanding of life science that a living individual is a living thing in the closed system in the universe, has prevailed in the 20th century and today.

There exist about three kinds of life on the earth. They are the prokaryote, eucaryote, and multicellular animal. A living thing takes in high-energy substance (nutrient) into the body from life media and utilizes the power obtained by decomposing it, in order to rotate the vortex of energy. Multicellular animals are susceptible to strong influence of energy operating against the outside of the living body. To the contrary, the prokaryote and eucaryote are hardly susceptible to the influence of gravity-energy or thermal energy(between 4℃〜40℃). The method to take into nutrients into a cell differs among the three. The prokaryote directly absorbs nutrients within media to rotate the vortex of life. The eucaryote directly absorbs nutrients within media, and in addition takes into as food vacuoles to digest and use. Directly absorbed substances are consumed by an organelle, and rotate respiratory metabolism. Mitochondria are aerobic prokaryotes parasitic on eukaryotes to perform oxidative phosphorylation and rotate TCA cycle. Multi-cellular organisms digest foods and the blood carries these nutrients into each cell. Each cell absorbs nutrients available directly from the blood that are used by mitochondria. In addition, they absorb nutrients by pinocytosis and consume by re-digesting within cells.

It is cell's life force, the power to overcome disease, namely, immune power that is the power to remodel rotating the vortex of energy by digesting things carried by the blood and by cellular respiration in multi-cellular animals. The impairment of mitochondria in charge of cellular respiration results in the impairment of production of collagen, cartilage and bone, and in addition disturbs remodelling. This is immune disease. In complete deficiency of Vitamin B , the function of mitochondria of the heart stops, which leads to instantaneous death. This is the so-called 'beriberi heart'.

Plants perform photosynthesis by having chloroplasts instead of mitochondria. An ascidian being the source of the vertebrate is an animal-plant organism that is a fork between animals and plants. An ascidian has the collagen-cartilaginous skeleton, intestinal respiratory branchia, and branchial heart characteristic of the animal, and moreover has the root of cellulose characteristic of the plant. This cellulose has been succeeded to kelp-like eggs in a primordial shark.

Evolution Of Life Viewed From The Intestine (1) –The Branchial Intestine What Is The Intestinal Respiratory Apparatus?

Higher multi-cellular organisms begin at the development of the intestine. The intestine develops at the starting point of evolution in the invertebrate or vertebrate. Multi-cellular animals take outside substance into the intestine. They digest and absorb it into the body to dissolve in the cell and take out energy that is provided for life activity. The dissolvement-method to take out energy from substance is referred to as energy metabolism. There are two kinds of dissolvement-methods: anaerobic glycolysis and aerobic combustive respiration. Dissolved substance is ordinarily hexose. Repeatedly, the living thing is referred to as the system in which it dissolves sugar and rotates energy vortex, and remodels old parts or cell of the body.

Replacement of an individual as a whole is done with reproduction in which an ovum and sperm combine with half genes. These cells are made of the intestinal duct and excreted from the intestine. That is to say, the intestine is the most important organ for multi-cellular animals. In a hag fish of Cyclostomata referred as to the living fossil of the most primitive vertebrate, scatter capsule-like eggs are scattered about 1 cm around the intestine, and there is no structure of the ovary. They are just superfluous nutrients absorbed from the intestine.

Relation between the Branchial Apparatus and Skeleton

Elucidation of the relation between osseous tissue and the cell-respiration system, can be obtainable by investigating morphologically the relationship between the branchial apparatus and skeleton.

A prochordate being the source of the vertebrate is an ascidian. An ascidian's cartilage exists only in dermal denticles (scale of the skin becoming the anlage of dermal tooth, scale and beasts' hair). An ascidian performs intestinal respiration in the branchia. It has the heart and pericardial cavity derived from the branchial gland, and has already every organ equipped with higher vertebrates at the cellular stand. However, since an ascidian is a monosegmental animal, birth of the vertebrate from it requires development of the multisegmental animal by gene duplication. While an ascidian is a monosegmental animal that has one heart in one branchia, an cyclosalpa in which an ascidian ties in a row has one branchia and heart in its respective segment (Fig. III-2).

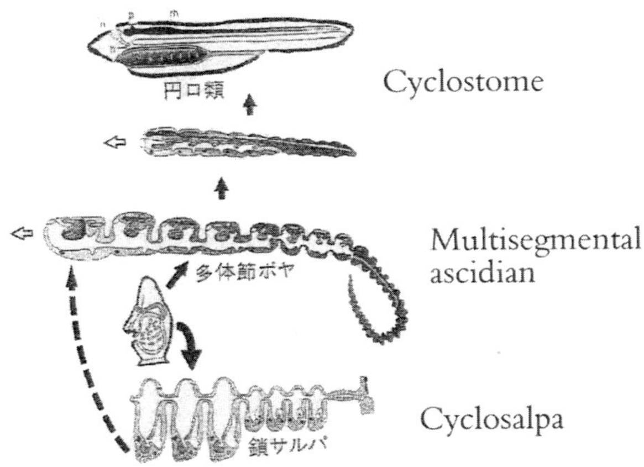

Fig. III-2. Evolution from an ascidian to a primordial fish

A lancet is similar to a multisegment cyclosalpa. A lancet has one heart in one branchia. The one heart in one branchia was originally the moving haemopoietic apparatus, namely, the source of the heart. Since the heart is thus derived from the haemopoietic apparatus of the branchia, a primordial shark is referred to as the branchial heart. The branchia was originally the moving blood vessel performing haemopoesis of erythrocyte with gene expression of mesenchymal cells triggered by oxygen in the water and by absorbed nutrients from the liver via the intestine.

The next stage's animal in evolution is Cyclostomata, or Agnatha, and it has neatly the cartilaginous jaw derived from branchial arches. Cartilaginous genuine teeth that succeeded the dermal denticle of an ascidian, grow in the jaw. The branchial gland of Cyclostomata moves amorphously just like the heart to produce erythrocytes and leucocytes, and is all covered with thin cartilaginous membranes. Since the heart is derived from the branchial gland that became one organ by combination of the right and left, as the heart is called the branchial heart till the stage of a shark, the heart exists naturally in the sack of cartilaginous bone. The cartilaginous sac surrounding the heart is the pericardial cavity. Since the cartilage is connected to the pectoral fin, the fin becoming the upper arm on the occasion of the vertebrate is in fact the system of intake of the branchia. The fin becoming the lower limbs is the excretion system of the branchia that assists excretion of urinary and reproductive substance (a kind of blood cell) and food debris. The next stage of evolution is spiny sharks among jawed vertebrates, and their descendants are sharks (cartilaginous fishes). Cyclostomes' cartilaginous sac covering the branchial gland except the pericardial cavity has been succeeded as the branchial cartilage in sharks.

Muscles of auditory ossicles, tongue and heart are derived from the branchial intestinal muscles

Among sharks, Chlamydoselachus retains ancient forms, and gill slits stand in six lines. All the other sharks have their first gill slits as ostia at the rear of their eyes, which are called aerial holes. They become the ears and their branchial glands become the internal ear. This means that in Devonian period when all the sharks except Chlamydoselachus were confined to brackish water, they were obliged to live in the water at high tide, and to perform air respiration at low tide just like Mutsugorou in Ariake Sea. If you consider that the cause of deviation from prototype is derived from materialistic change of life media (water, oxygen, and air) under the function of gravity, and change of biomechanics (energy), you can surmise accurately the changes of past lives from the standpoint of the changes of forms.

In the other sharks except Chlamydoselachus, the first gill slit becomes the ear hole, of which branchial gland becomes the inner ear, and branchial arch cartilage becomes auditory ossicles cartilage. Originally, they had the centralized system of lateral line in old jaw bone belonging to the first branchial arch where they heard sound. The second branchial gland to the sixth becomes the haemopoietic apparatus of venous sinus (large vein from the heart) leading to the branchial heart. The second branchial arch cartilage to the sixth gathers at the centre of the mouth in the inside of the mandible just like fan-shaped bone to mould the immovable tongue (Fig. III-3).

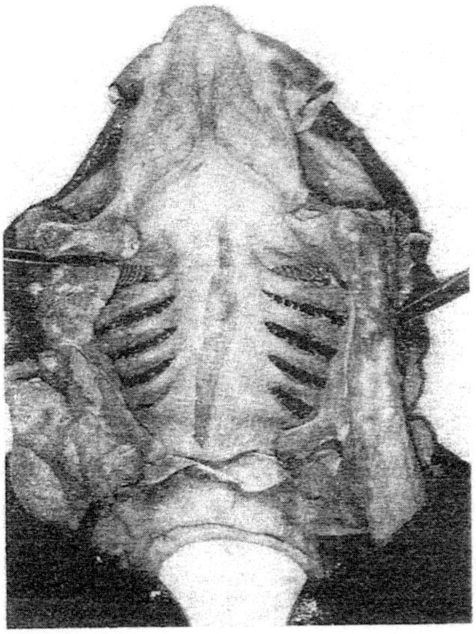

Fig. III-3. The teeth of Heterodontus and the immovable tongue.

To this branchial arch is attached the branchial muscle. This form of the branchial arch cartilage coincides perfectly with the one of an aquatic axolotl. When we get axolotl with neoteny ashore, its external gills performing fan-shaped movement shrink to be missing, and gill slits close, and epidermises really fuse. The fan-shaped gill four branchial arch cartilages which gather in the mandible to form the immovable tongue just like the case of a shark. With the disappearance of external gill, branchial arch cartilages too shrink, and finally are connected in a row at the rout of tongue to form a hyoid bone. The immovable tongue is composed of visceral muscles moving the branchia, and the bundle of the root of the branchial arch muscle gathering fan-shaped bones is of tongue-shape. The tongue moves each branchia like fanning as the pivot of the fan. After the moving external branchia involutes on being got ashored to disappear, fan-shaped bones with tongue-shape almost disappear by the Use and Disuse Law, and the tongue is only composed of branchial arch muscles at which root the hyoid bone sustains tongue muscles. Since the heart is originally derived from branchial arch muscles, the tongue of a shark contains as far as the caudal bottom of the pericardial cavity surrounding the heart (Fig. III-4).

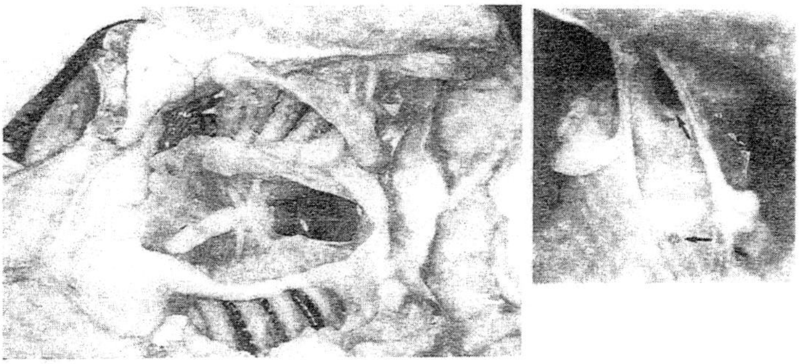

Fig. III-4. Left: the primordial cavity of Heterodontus (the heart visible in the inside). Right: a pneumatic hole formed in the pericardial cavity by landing (an arrow).the right of the photograph is the heart, and the left is the liver.

Heterodontus Becoming The Mammal

Sharks are separated from ancient Cladoselache into Triakis-type and Heterodontus type.

Only Heterodontus differs in all of the cranium, branchial apparatus, skeleton, and dermal tooth from other sharks. Heterodontus's shapes of the cranium and external nose, the mandible and tooth, and the branchi, the heart and the branchial arch differ from those of Triakis, and are similar to those of Cyclostomata. The heart of Triakis is small, and their pericardial cavity is completely covered with cartilages just like that of Cyclostomata. The heart of Heterodontus is extremely large to have cartilages on both dorsal and abdominal sides, which open widely at the entrance of vessels coming from the heart to the branchi. When the aerial pouch derived from the branchial sac moves backwards to enter into those portions, those portions become the mediastrinum (the partition separating halfway the both lungs existing only in the chest-portion of the vertebrate, which contain the heart, branchi, oesophagus, aorta, thymus gland, lymph nodes and so on). To the contrary, the aerial pouch (the lung) of Triakis's branchi moves backwards to the pelvis of the abdominal cavity in parallel with the esophagus, being close to the pericardial cavity surrounded by cartilages. That's why the lungs of amphibious, reptiles and birds reach their intestines and their lungs and other internal organs are mixed up in their abdomens.

The reason why the mammalian chest contains the thymus gland, lung, heart and esophagus alone, is that the lung being the branchial aerial sac moves backward to enter between the vessels coming to the branchi from Heterodontus's esophagus and heart. Two cartilages derived from the pericardial cavity existing in the back and belly become the sternum and tracheal cartilage. (Fig.Ⅲ-5)

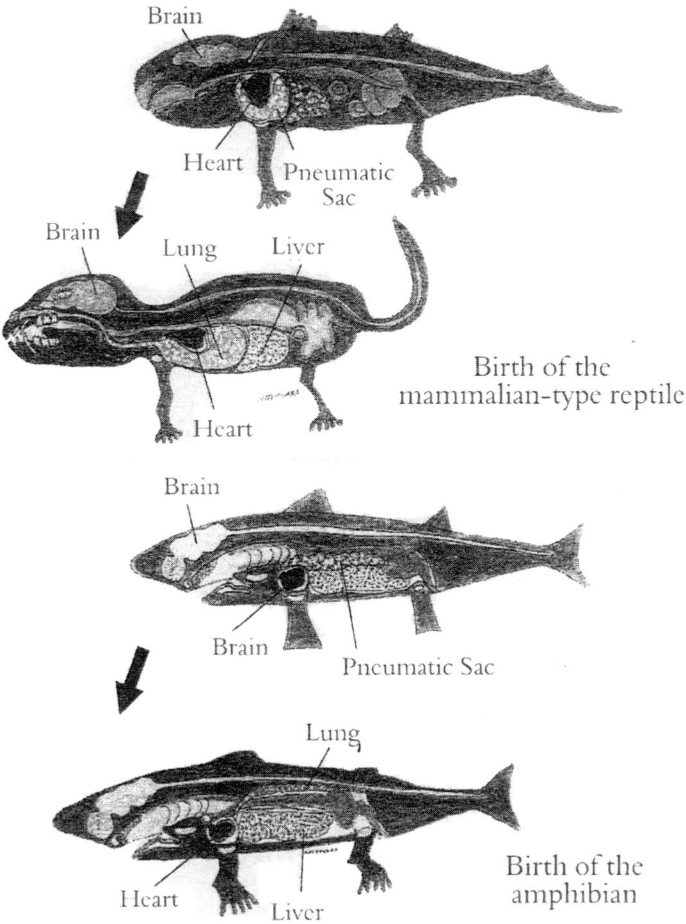

Fig. III-5. From Heterodontus to the mammalian type (upper illustration). From Triakis to an amphibian (lower illustration).

While in axolotl the cartilages of the chest fin adjacent to the pericardial cavity impart thin board-appearance consisting of the two sheets on either side like a stirrup on the abdominal side of the heart, in mammals they pushed up toward the head to become the clavicle because the mammalian lungs enter into the pericardial cavity. That's why the phrenic nerve in the caudal bottom of the pericardial cavity innervates the supraclavicular nerve. At the landing drama of the vertebrate, the branchial gland of the branchial apparatus, and the branchial cartilage, namely, the branchial respiratory system changes radically. Since both the branchia and bone are characteristic organs of the vertebrate, elucidation of the law of their transformation will clarify the causes of evolution. You can easily program two experiments for it.

After landing an axolotl followed by Heterodontus and Triakis, their structural changes were observed. These two experiments have clarified that the development of the lung has been separated into the mammalian type and the other two types, and both the development of the lung and transformation of the branchial apparatus turned out to be metaplasia. The skin of axolotl starts performing dermal respiration after artificial landing. If you want to know what dermal respiration is, all you have to do is to observe the tissue specimens. The skin after landing is developing promptly erythrocyte-haemopoesis in the subcutaneous tissue. This is metaplasia-phenomenon caused by gene-expression of undifferentiated mesenchymal cells with oxygen being a substance with mass. In this situation, the flimsy large heart in the water reduces by a third in size, and the vessel and nerve nourishing the heart begin to develop. This transformation of the heart has been caused by gene expression of cells producing the heart and blood vessels with the increasing streaming potential accompanied by the increasing blood pressure caused by the increase of gravitational stimuli. This is metaplasis caused by energy having no mass.

Transformation of the Branchial Apparatus by Metaplasia

At the landing of Heterodontus, the second branchial gland to the fifth closes, and only the sixth opens. The second branchial gland becomes Waldeyer's tonsillar lymph ring, namely, the haemopoietic apparatus producing leucocytes, the third and fourth becoming the thymal gland, and the fifth becoming the ultimobranchial body. The sixth becomes the lung by metaplasia caused by oxygen. This transformation, i.e. metamorphosis of the branchial gland is also metaplasia triggered by substances (material and energy). The diminution, fusion and ossification of the branchial arch cartilage is differentiation- induction-phenomena of osteoblasts, which is also metaplasia, occurring with haemopoietic nests of undifferentiated mesenchymal cells caused by the increase of streaming potential, namely, the increase of gravitational function (6 times) accompanied by the change from the environment in the water to that on the land. In this situation, when oxygen increases by 20-30 times, cell respiration is activated. The oxygen in charge of cell respiration is mitochondria. Insomuch as mitochondria are procaryotae like chloroplast in the plant, they have a set of components such as specific genes, protein-synthesis-system, DNA-RNA polymerase and the like. They dissolve glucose to rotate TCA cycle and produce ATP being pyrophosphate that is high-energy substance (high energy substance mediating the transmission of energy). This is oxidative phosphorylation in which at mitochondria calcium and phosphoric acid are condensed to release hydroxyapatite. Cartilages and bones provide the substances for glycolysis to produce thiolester and ATP, and respiration, but also are products by metabolism and rotation of these energies. Thus, the branchia in charge of energy metabolism is covered with cartilages from the starting point of primitive cyclostome which become the branchial arch at the next stage, functioning as the intake of oxygen at the external respiration as the fan-movement of the branchia.

At the landing drama of the second revolution of evolution, when the function of mitochondria activates by a leap of cell-respiration, these branchial arch ossifies. The animal is characterized by active movement. The skeletons to sustain its operating system are made of substances to support energy-metabolism, and at the same time have been the myelopoietic organ, or the generator of blood cells in charge of cell respiration. I previously mentioned the substance to determine the vertebrate was bone and its characteristic organ was the branchia in charge of intestinal respiration. These two essential organs have been perfectly connected through mitochondria. By this, the riddle of the 21st century's life science has been completely solved.

Evolution Of Life Viewed From The Intestine (2) The Abdominal And Excretory Intestine Origin Of Life "Intestine"

The original organ of higher living things is the intestine. Around the whole of the intestine resides your heart, where the origin of desire for fortune, fame, sex, food and sleep exists. The intestines are roughly divided into three. The intestine in charge of respiration is the branchial intestine, and the pitch-dark intestine begins after the throat. This is the abdominal intestine, which meanders seven-eight fold to become the urinary, reproductive and anal intestine. The author calls this the excretory intestine in contrast with the branchial intestine. It refers to the intestine for excretion.

In rhythmic exercise of the muscular movement of the branchial intestine dwell pity, affection and emotion. In the rhythmic movement of the smooth muscles of pitch-dark intestines wriggles the self called the mind of the desire for survival with coiling itself up. The abdominal intestine has an itch under accumulating conditions. The mind of the excretory intestine is usually called carnal desires, which have been looked down in the way of doing things in civilized society. While the true natures that the whole of the intestines from the mouth to the anus are the desires for property, fame, sensual pleasure, eating, and sleep, the desires that the branchial, abdominal and excretory intestine represent are subtly different, respectively.

The Intestine of Heart "The Branchial Intestine"

In mammals, the organs stemmed from the branchial intestine represent emotion and spirit, 'heart leaping', buoyant and cheerful', 'filled with joyful expectation', and 'heart beating', represent high spirits caused by the throbbing lungs and heart. In negative expressions, there are a foreboding, heart torn to pieces, heartrending, an oppressive feeling in one's chest, choking, suffocating, horrified to heart, deeply worried, grieved, and anguish. Their sorrow conditions in the lungs and heart are represented with tears, nasal mucus, and voice by the muscles of the face, the eyes, nose, and vocal code, which are all the muscles and its attached organs derived from the branchial intestine. Heart is the feeling derived from the internal organs such as happy, sad, angry. This heart dwells in the chest and abdomen.

The branchial intestine exists in the cephalon, and makes cephalad movement to react with coming across incidents. As a result, it represents the mind with the throbbing heart or branchia (lungs) by vasomotor reflex of the sympathetic nerve. Since in mammals, the sixth branchial gland moves backward by the law of inertia and air pressure to enter into the pericardial cavity surrounding the heart existing at the base of the tongue, the heart and lungs, and tongue have become one branchial apparatus. Pathological anatomy recognizes the tongue, mediastinum, and the heart and lungs as one block. As previously mentioned, a simultaneous transplant of the heart and lungs replaces one's heart and mind with those of the donor. Needles to say, it is quite natural since it heart being the main theme of this book exists hereabouts. Since the heart is the muscle of the branchial intestine and the lungs are larger than the liver derived from the intestines, it is quite natural that transplant of the intestine brings about these results. Le

Douarin's self and nonself identity did not exists in the major histocompatibility antigen, but identity did exist in the intestines themselves. Terrible to say, transplant of the heart and lungs of an erotomania makes really the recipient an erotomania. This kind of absurd transplant should be stopped as soon as possible.

Self Resides In 'The Abdominal Intestine'

The pitch-dark abdominal intestine follows the esophagus. Since self really resides in the abdomen, old worriers would commit hara-kiri (cutting their bellies) when they had failed to achieve self-fulfilment. If they didn't commit hara-kiri, their anger would not subside. Self is the desire equipped with living individuals based on intrinsic five desires, namely, one's true character, and impudent visceral sensation residing in the intestine. The word 'abdomen, belly , or stomach' has been used as the word representing one's undistinguished character connected directly with self as follows: angry, black-hearted, talk heart-to heart, find out what is in a person's mind, determined to do, can't contain one's anger, rage, cant stomach one's- , boiling with rage, heartrending, big-hearted, to be perfectly honest, and so on. Real anger wells up from the bottom of the belly (the intestine).

The Excretory Intestine Wanting Excretion and Remodelling Of Cells

The excretory is the intestine representing desire for excretion among five desires of the whole intestines. What are excreted, are metabolite, food debris, and superfluous nutrients that are reproductive cells and so on in the remodelling system as the whole of individuals. Nutrients absorbed from the intestines are all taken into blood and its cells, and once again are digested and absorbed at cellular levels. The metabolites remodel the body cells and parts, and old materials are replaced by new ones. And the blood carries their metabolites into the kidneys, which are excreted as urine. It is leucocytes that perform diligently this remodelling. In man, one billion cells among 60 billion nes are replaced at one night. This is currently called metabolism, of which realities have been known. What has become of superfluous nutrients after metabolizing? Superfluous nutrients are converted into fats to be stored, and part of them becomes an ovum, or sperm being a kind of blood cells. The amphibian has an ovum and sperm at the head of a lump of fats followed by the kidneys. The kidneys are originally haemopoietic nests of the mesodermal branchial system as pronephros, and became mesonephros and metanephros by going backward, where erythropoietin (protein to differentiate and induce blood cells, cytokine) differentiates and induces superfluous nutrients into a sperm or an ovum by gene expression. Why is it known that the kidneys were part of the branchi? The reason is that all the organs innervated by the vagus nerve coming from the medulla oblongata called the branchial brain had been the branchial apparatus in ancient times. However, the bladder, testis, uterus, ovary, rectum, and anus (primitive-type reproductive organ) except the kidneys in the urogenital organs, have been governed by the excretory brain. Only the adrenal cortex has been governed by the parasympathetic nerve (recognized as the sympathetic nerve in the conventional anatomy) of the spinal cord of the abdominal brain.

Every cell of the body remodels through blood. Metabolites are carried by blood, and transformed into urine by the kidneys that are a peculiar haemopoietic organ. On the other hand, superfluous nutrients also undergo meiosis at the organ connected with the kidneys and become an ovum or a sperm being peculiar blood, which is excreted by the same duct as urine. Urine and reproductive substance are perfectly equivalent for the living bodies. Namely, production of a sperm, or an ovum and that of urine have been intimately related with palpitation by the heart or breath by the lungs. And these sperm, ovum and urine are excreted from the excretory intestine. Accumulation of them lets the excretory intestine being itchy. Then, spiritual activities or heart functions being in charge of the branchial intestine turn out to stop, and thinking power paralyzes in most occasions. The excretory brain resides in sacral region of spinal cord, and governs the rectum, urinary bladder, testis, ovary, uterus, vagina, penis, and testicles. Since every itch of the excretory intestine, regardless of desireurge, defecation desire, menses, ovulation, projectile ejaculation, fecundation, delivery, is the desire connected directly or indirectly with remodelling of the whole individual, it is the most intense desire among five desires accompanied by unequalled pleasure. It is the nearest mind to the purpose of living.

The Intestinal Brain' And Nerve

The brain governing pitch-dark intestines is the intestinal brain. The brain known as the branchial brain (oral brain) and the excretory brain (anal brain), is composed of the combined nerve bundle of nerves of the skin (ectoderm) in the exit and entrance of the intestine and those in the intestinal membrane (endoderm). The nerve is originally the system of muscles.

The nerve transmits membrane potential in the epithelial membranes and its function is to move mesodermal muscles. Accordingly, the double rule of the nerve in the skin (ectoderm) and that in the intestinal membrane (mesoderm) exists in the intestines of vertebrates. At the mouth and anal region, there are paraneurons in which ectoderm and mesoderm are combined, and are brains in charge of the most intrinsic living system for life activities and the reproductive system. Since only the intestinal brain does not have neurons that show the conditions of the inside of the intestine on the skin, the existence of the intestinal brain was not clear. These three brains have already existed since the days of ascidians. An ascidian (protochordate) shows the prototype of vertebrates, and at the initial source of this phylum the autonomic nerve ruling the internal organs forms the brain. Until the stage of a shark, the brains have been produced by the parasympathetic nerves in the internal organ system and by the extrapyramidal system of the parietal system alone. Thus, of the three brains existing since ancient animals, the nerve systems except the parietal system are all composed of the parasympathetic nerve system alone.

There is a book “Chou wa kangaeru” (Intestines think), but what originally should be thought is the skin-parietal system. Since the cerebrospinal nerve system is a specialized skin, “Intestine also think” is proper. Though preganglionic fibres have been recognized before as the sympathetic nerve in spite of cholinergic activity (the parasympathetic nerve is effected by acetylcholine, and the sympathetic nerve is effected by adrenalin, or noradrenalin), they should be recognized as the parasympathetic nerve, since sharks have no sympathetic nerve. Sharks have no pyramidal system and no sympathetic nerve. And their every movement is reflex movement by the extrapyramidal system.

Food and reproduction alone make sharks move, and they rest during other times. Since moving muscles by feeling anything is the start of thinking in fact, sharks also are thinking of eating and reproducing alone. It is peristalsis of the intestinal tracts in fact that evoke these two kinds of behaviour. They think of eating when hungry, and think of reproducing when filled with food and being itched, which is the start of thinking. The pitch dark intestines show only impudent self existence, because they don't have any paraneurons at all in the somatic system. The face expresses emotion or mind, and complexion or behaviour expresses sexually aroused state, because both the branchial and excretory intestine have the eyes, touch sense and genital organs of paraneurons that connect the somatic system with the brain showing the mind in the intestine toward the outer world and the sensory organs. Since the abdominal intestines don't have them, probing a person's real intentions is impossible. The reason to commit hara-kiri in order to prove one's innocence is that one has no neuron to show one's intentions in the belly. Therefore, one is obliged to commit hara-kiri to prove one's innocence in exchange for one's life.

When we view the intestines by three divisions, we find out that the branchial intestines are in charge of heart, spirit, thought, and about 90 percent of all intestinal ducts. The kidney system producing reproductive cells is also innervated by the branchial intestine. The excretory intestine represents itchy mind caused by produced ovaries and sperms with the body. Spiritual activities, especially cultured activities are self-fulfilment and reproduction of spiritual self on an enlarged scale. Therefore, they are quite near to reproduction of life on an enlarged scale. Desire for fame is very similar to sex drive. The desires in the intestines in the belly are property, food and sleep, and are desires for existence. The excretory intestine is the so-called sexual desire that is hardest to control.

The branchial intestine and its appendages such as the eyes, nose, and auditory static organs, are transformed into the complex system ranging from the head, neck, breast, diaphragm, and kidneys to reproductive intestinal organs. The liver, spleen, pancreas, and most of stomach and intestines are the abdominal intestine that in fact is derived originally from the digestive part of the branchial intestine. The cells becoming all anlages of these organs are scattered in the coelenteron of a Bugula at the age before an ascidian, and a hydra at the further former age as well.

While these various kinds of cells gather to produce organs, the epithelial system is controlled with ion-channels by membrane potential, and the mesenchymal system is controlled by streaming potential. The reason why the abdominal intestine is double-innervated by the vagus nerve and spinal nerve, is that the branchial intestine system established at the monosegmental ascidian age has been functionally retracted by cephalad movement after segmentation. The branchial apparatus gathers at the head side by cephalad movement, and the stomach and intestine innervated by the branchial nerve move backward to the anal side. And when every segmental intestine is in sequence, it becomes the helical intestine. When cephalad movement after territorialisation at this stage speeds up, the internal organs and brain are subject to enervation by the sympathetic nerve. One frame of the helical intestine also connects with each other by the same reason that the branchi connects in consecutive order. During the period when running water from the mouth contains oxygen, only ascidian's branchi remains, and the digestive ducts diminish by the law of Use and Disuse. When oxygen becomes vacant, an ascidian's intestine in charge of food-digestion, in turn functions, and the branchi disappears, thus the helical intestine being in succession.

This intestine becomes the pitch-dark abdominal one by dual innervation of the vagal nerve and spinal parasympathetic one. There exist lumps of emotions consisting of fortune, fame, sensual pleasure, appetite, and sleep with coiling themselves up in the rhythmic exercise by the smooth muscles of the swirling helical intestines. A portion of them operates in the face, neck, and breast derived from the branchial intestine, and another portion engages the function of carnal desires as the excretory desire of the excretory intestine. Because mentality is incompatible with low passions, famous people are often obliged to ruin themselves by the failure of the conflicts between the branchial brain and excretory one. However, without sound and well-functioned excretory brain, descendants would not be handed down. Involvement of the spinal parasympathetic nerve with this helical intestine is due to the development of nutritive capillaries. Accordingly, after the oral cavity and anus were bipolarized by cephalad movement, namely, after gnathostome following cyclostome were established, this constitution was completed.

The reason why the abdominal intestines of sharks were double-innervated by the parasympathetic nerve is as follows: because of rapid physical exercises, rapid cephalad movement, and long periods of cephalad movement (about 100 million years), the gastro intestine moved backward to the anal side, where blood vessels accompanying with the parasympathetic nerve from the spinal cord have gotten long. This is the gravity-corresponding evolutionary study. Since shapes change by body-structure, law of inertia by gravity function, and the speed and duration of cephalad movement, time, function energy, or functioning substances (oxygen, water, air, etc) could be found out by calculating backwards in terms of shapes. The helical intestines have become bulging and narrow segments of the mammalian intestines.

Life Evolution Viewed From The Nerve And Muscle (1) Dorsal Muscles And The Pyramidal System Source Of Thought And Spirit In The Somatic System

There is a saying, "The belly can not be replaced with the back" in Japan, meaning "There is no other way out". Rhythmic exercise by the dorsal muscles develops spirit and thought, and one by peristalsis in the intestinal ducts develops heart and mind. For the most part, spirit and thought are incompatible with instincts such as property, fame, sex, appetite, and sleep caused by intestinal rhythmic exercise becoming the source of heart. In the proverb of "The belly can not be replaced with the back.", the back refers to the parietal muscle system, and the belly refers to the intestinal visceral system. It is the mouth and anus that connect the back with the belly: the mouth is entrance of the intestines derived from visceral muscles having skeleton moved by the power of will, and the anus is exit of them without skeleton.

The dorsal muscles of mammals are quite different from those of primitive-type sharks. Shark's muscles of cartilaginous fish being a primitive vertebrate do not involve the pyramidal system, and as to the autonomic nervous system, they have only the parasympathetic nervous system with no sympathetic nervous system. At the chance of landing in the second revolution of evolution, when gravity function multiplies by six times, and the living medium changes from the water to the air that is a thousand as weigh as the water, and oxygen content changes from 0.7% to 21%, cell respiration of every tissue and organ in the body activates inevitably just by corresponding with the environment, and induction of blood vessels occurs. The muscle is a neural functional organ. Regardless of sensible (the parietal system) or non-sensible (the visceral system), when information is transmitted to the centre by afferent nerves, electric potential develops at the central nerve nucleus, so that information is transmitted by efferent nerves to move muscles

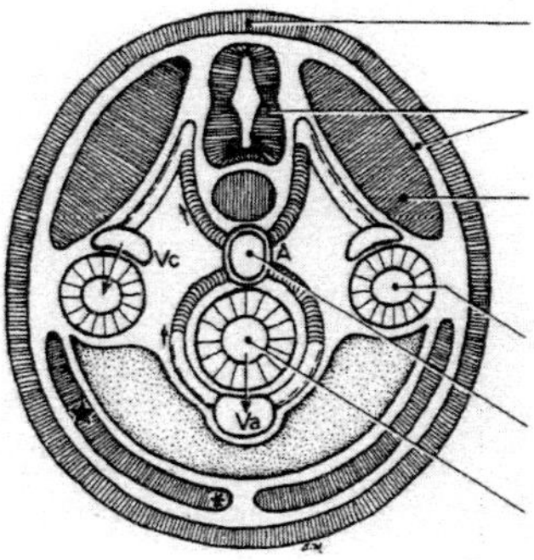

Fig. III-6. The somatic and visceral system of the vertebrate (Original by Shigeo Miki, [the work of the internal organ and the heart of children] from Tsukiji shokann) .

After the completion of the oral and anal bipolarization in the parasympathetic nerve, the parietal muscle motor system grew by leaps and bounds on landing to cause the development of the sympathetic nerve and to cause the development of the brain motor nerve in the pyramidal system. As a result, respiration and energy metabolism of the glycolytic system grew by leaps and bounds to cause the rapid increase of heat, resulting in homeothermism of cool-blooded animals. At the same time the tongue or the limbs (fins) come to be moved by the power of will.

While the shark being the primordial vertebrate is a descent of Acanthodii, its tongue is before landing a gathered portion of the branchial arch of the branchia and the muscles, and has a form the tongue already, moving slightly. On observing the sagittal section of either Triakis or Heterodontus, you'll find the heart surrounded by the pericardial cavity at the backmost portion of the rhythmic branchial arch muscle's row composing the slightly movable tongue. Surprisingly, the source of thought and spirit has resided at the pyramidal system of the somatic muscle system represented by dorsal muscles. From the first, existence of the safe and sound pyramidal system is essential for spiritual activities. "A sound mind in a sound body" refers to it, and the source of mind and thought exists in the parietal muscle system, with which muscles the parietal brain or the cerebral neocortex is conjugated. Rigidity of dorsal muscles by brainwork is due to this. Brainwork also needs to introduce rhythmic exercise of muscles tuning to respiration. For this, it is essential that paleocortex in the limbic system or the pyramidal system in the parietal system of the cerebellum should be safe and sound.

The relationship between the development of the jaw and oral cavity and that of the sympathetic nerve and the pyramidal system necessitates the formation of mastication. When Heterodontus is confined to waterfront and wriggles with agony on landing, a large volume of oxygen that becomes thirty-fold than before enters its body.

Struggling moves muscles, and cell respiration of mitochondria activates with the result of rapid consumption of glucose. Since Heterodontus by nature used to masticate well and eat shrimps or shells, they would supplement nutrients by masticating seaweed, small fish or shells well. If glucose is not supplemented, new birth of blood vessels will not occur whatever oxygen increases. Thus, cell respiration that is mastication of food being energy-source, occurs in synchronization with development of the lungs being the air respiration system.

Thought and Spiritual Activities Begin at Rhythmic Exercise

Since the human tongue is an important organ in charge of expression of mind and spirit, it is essential to observe development phylogenically. It has been recognized that the hypoglossal nerve does not exist in sharks but in mammals, and became the twelfth cranial nerve, because the spinal cord has been taken into the brain. However, it is clearly incorrect that the shark's tongue belongs to the somatic skeletal muscles by the following reasons: nucleus of the hypoglossal nerve exists at the same level of that of the vagus nerve, all sharks' tongue muscles with slight movement being composed by the branchial arch muscles, and all the sensory nerves innervating the tongue muscle being the branchial arch nerve. Smooth movement of the tongue is due to retraction of the branchial arch cartilage accompanied by retraction of the branchial intestine after terrestrialisation.

With disappearance of the rhythmic movement of the branchi by becoming proficient in the use of pulmonary respiration, when the branchial arch cartilage retracts from the tongue being an aggregate of branchial intestinal muscles and branchial arch to become a hyoid bone, the tongue composed of branchial intestinal muscles released from skeleton begins to move. Since this is coincident with the development of the sympathetic nerve and pyramidal tract system, the tongue brings about rapid increase of the nerve cells in the cerebral cortex motor area that develop accompanied by rhythmic movement created by will characteristic of somatic striated muscles. This is also due to the Use and Disuse law. At this point, the masticatory muscles, expression muscles, swallowing muscles, and phonetic muscles that are derived from the branchial muscles governed by the cerebral nerves of the pyramidal system, are all transformed into the muscles governed by the will of the somatic system.

Mind and thought begin at the stage when the cerebral pyramidal system and sympathetic nerve system have developed in the somatic system of the vehicle carrying freely the intestinal ducts, and the nutrient blood vessel system have entered into every intestinal organ (from the heart, intestinal internal organ to the brain). Until then neither a shark nor an aquatic axolotl does not have the coronary artery, and does not have the nutrient artery feeding the intestine. The sympathetic nerve enters into the internal organ to the brain together with the nutrient vessels to inform the internal organ or brain of the somatic conditions.

When the brain's protruding organs such as the eye, nose, touch sensation, or taste sensation, perceive outer occurrence, mind first of all informs the somatic skeletal motor organs of it, resulting in escaping from terror or being drawn to the opposite sex. Moving for foods and at the same time informing the heart, intestinal visceral system, or brain via vessels of it, becoming buoyant and cheerful, or being seized with terror for encountered conditions, one measures escapes or safe places by the width of hands or legs. At this point, thinking begins, and thought and mind begin. Mind is the representative organ expressing the effects of the branchial brain, and it resides in the somatic muscles such as the face muscles composed of the striated muscles in the pyramidal system and visceral muscles, and muscles of the back or limbs.

The activities of mind and thought develop by rhythmic movements of those muscle groups, by adding functions of the somatic motor system and pyramidal system to the muscles in the excretory intestine and excretory system that support the basic constitution of eating, respiration, digestion and reproduction in the primordial branchial intestines having lived with the parasympathetic nerve and pyramidal system alone. By the synchronized rhythmic exercise with respiration called "language" acquired by the cooperative functions of the facial expression muscles, tongue muscles and cervical muscle group in the branchial intestinal region, the activities of thought and mind have developed greatly to give birth to mankind. Language has been diverted from rhythmic movement of muscle groups functioning by peristalsis of the visceral cranium being eating or mastication to rhythm at the thought level in the sympathetic nerve system and pyramidal system.

Life - Evolution Viewed from Nerve and Muscle The Pyramidal System Development of the Nervous System

The beginning of the vertebrate of multi-cellular system is a hydra of the coelenterara that respire by the skin. This is followed by polarization of respiratory cells, that is, respiration by alae in Bryozoa, and moreover when respiratory cells are taken into the intestines, an ascidian of Prochordata is born. Differentiation into respiratory cells is owing to metaplasia in which oxygen concentration in the water triggers.

Both a hydra and an ascidian are a sac and a duct made from intestinal muscles. The nerve is the system of the muscle; it is the system to digest food by peristalsis, or to move the body according to desires by intestinal ducts. The basic constitution of the vertebrate begins at a sac. Part of it is broken to become a tube, which constitutes the three layers of the ectodermal, endodermal, and mesodermal epithelium. The basics of the nerve are transmission of electric potential by connection of epithelial cells, namely, by membrane potential. The primordial vertebrate has two kinds of the nerve system: one is that of the intestinal visceral epithelium, and the other is one of the somatic system connected by subcutaneous cells of the ectodermal system. The former is the nerve system of smooth muscles by peristalsis to form the visceral brain of the parasympathetic nerve system. The latter is the cerebral nerve of the somatic system, which is composed of only the extrapyramidal system in the primordial type. In the extrapyramidal system, the left cerebral nerve governs the muscles of the left side of the body.

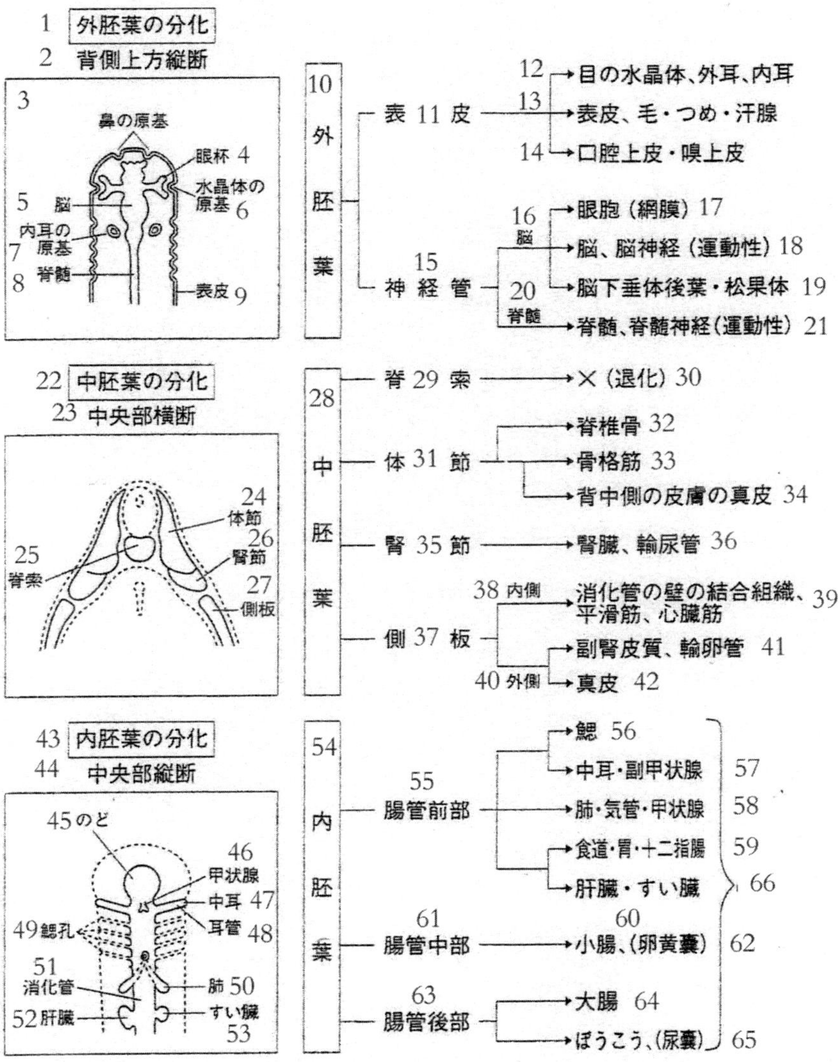

Fig III-7. Differentiation of the ectoderm, mesoderm and endoderm (The initial stage of ontogeny)

1 Differenciation of the ectoderm. 2 Dorsal upper longitudinal section. 3 Anlage of the nose. 4 Eyecup. 5Brain. 6 Anlage of the crystalline lens. 7 Anlage of the inner ear. 8 Spinal cord. 9 Epidermis. 10 Ectoderm. 11 Epidermis. 12 Crystalline of the eye, outer ear, inner ear 13 Epidermis, hair, nail, sweat gland. 14 Oral epithelium, olfactory epithelium. 15 Neural tube. 16Brain. 17 Ocular vesicle (retina). 18 Brain, cranial nerve (motor). 19 The pituitary gland posterior, corpus pineale. 20 Spinal cord. 21 Spinal cord, spinal nerve (motor). 22

Differenciation of the mesoderm. 23 Central cross section. 24 Somite. 25 Notochord. 26 Nephrotome. 27 Lateral mesoderm. 28 Mesoderm. 29 Notochord. 30 Degradation. 31 Somite. 32 Vertebra. 33 Skeletal muscle. 34 Dorsal side corium of the skin. 35 Nephrotome. 36 Kidney, ureter. 37 Lateral mesoderm. 38 Medial side. 39 Connective tissue of the digestive tract wall, smooth muscle, heart muscle. 40 Lateral side. 41 Adrenal cortex, Oviduct. 42 Corium. 43 Differenciation of the endoderm. 44 Central longitudinal section. 45 Throat. 46 Thyroid gland. 47 Middle ear. 48 Eustachian tube. 49 Gill-slit. 50 Lung. 51 Digestive tract. 52 Liver. 53 Pancreas. 54 Endoderm. 55 Front portion of the intestine. 56 Gill. 57 Middle ear, parathyroid gland. 58 Lung, trachea, thyroid gland. 59 Esophagus, stomach, duodenum. 60 Liver pancreas. 61 Middle part of the intestine. 62 Small intestine (York stak). 63 Rear part of the intestine. 64 Large intestine. 65 Urinary bladder (Allantonic sac). 66 Epithelium of… .

The animal is characterized by making its body move. However, it does not move at random, but in accordance with desire. Move by feeling is a start of impression, or a start of mind. What does what feel? The answer is that the intestinal ducts let the body move in search for food and reproduction. Accordingly, the nerve has an afferent (toward the brain) sensory nerve, and an efferent (exiting from the brain) motor nerve. There are neuron and paraneuron in nerve cells. Both are the system to transmit information (signal) by membrane potential at the vesicle of synapse. Roughly speaking, neuron transmits information to nerve cells and muscle cells, and paraneuron works as an effector to excrete mucus and so on.

A hydra being the beginning of multi-cellular organisms is composed of a sac of the intestine, and the beginning of movement of the intestine is the move by waves in the water. Cell groups having the same gene converge to constitute three layers of endoderm, mesoderm, and ectoderm. At the stage of an ascidian, muscles derived from are the intermediate type between smooth muscle and striated one, just like our heart muscle. The nerve system is as a rule composed of connection of the epithelium of ectoderm and endoderm, namely, work of membrane potential at the cell membrane. In the skeletal system of mesoderm, the trigger of cell genes being pulled by streaming potential caused by the flow of humour, differentiation and induction of collagen, cartilage, bone, or blood cells are controlled.

The Sympathetic Nerve And Pyramidal System Nerve

In the process of evolution of the vertebrate, it is the tongue that is the organ of the muscle system changing most remarkably. It is the sympathetic nerve and the pyramidal system nerve of the cerebrum that develop most remarkably among the nerves.

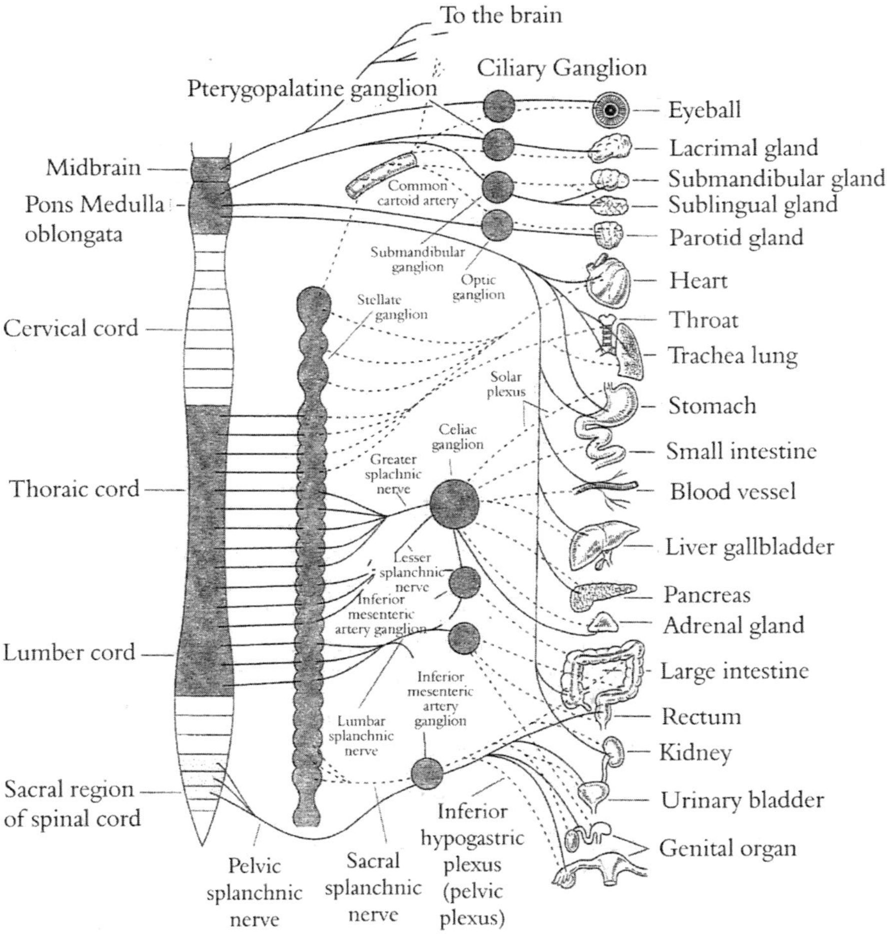

Fig. III-8. The parasympathetic nervous system and the sympathetic nervous system. The solid line shows the parasympathetic nervous system. The dotted line shows the sympathetic nervous system. One organ is subject to the control of the both nervous systems. The kidney turns out to be only one organ to be influenced by the oral brain (branchial brain) and the anal brain (excretory brain). [Original by Nishihara]

The pyramidal nerve system exits from motor area of the cerebrum, and its right and left intersect at the medulla oblongata. Since its course shows a pyramidal form, it is motor nerve called the pyramidal system. The function of the cerebrum that governs muscles according to thought, is to govern movable muscles by will. This nerve develops together with the sympathetic nerve when an organism landed and wriggled with oxygen increasing. There exist no these muscles in the pyramidal type's shark, since the shark does not have movable muscles governed by its thought. All motions of the shark are to be performed reflexively by the muscles of the extrapyramidal system according to the motion of the smooth muscles of the intestine and the work of the parasympathetic nerve system. The muscle is the effector of the nerve and the functional organ. Therefore, when the muscle changes in response to the change of function, namely, the animal's way of movement, the nerve develops in accordance with it. What do the development of the sympathetic nerve system and that of the pyramidal system respond to? This also responds to the second revolution in evolution of vertebrates. On the landing dram, oxygen increases rapidly as compared to in the water, and gravity seemingly increases by six times. Life medium changes from water of which thermal capacity is the largest, namely, specific heat 1, to air of which thermal capacity is extremely small; the weight of life medium decreases by a thousandth.

Environmental Factors in Evolution and Metaplasia

These biomechanical changes have been vaguely called environmental factors. While an archaeo-shark was writhing in search for water in response to these changes, it survived at the shallow pool that was left by repeated 'ebb and flow of the tide', causing automatic changes of the blood vessels of cartilage, gland or muscle of the branchi with the same genetic characters. The heart of a shark (and an aquatic axolotl) is an hemopoietic organ and has not blood vessels to nourish the heart. With the development of blood vessels, the sympathetic nerve develops as motor nerve. I want you to remember that all of a great number of cells to produce individuals of the vertebrate have the same genes. It is blood that delivers equally nutrients including oxygen into the bodily cells. While writhing after landing, a shark's blood pressure rises from 15 millimetre of mercury in the water to 30.

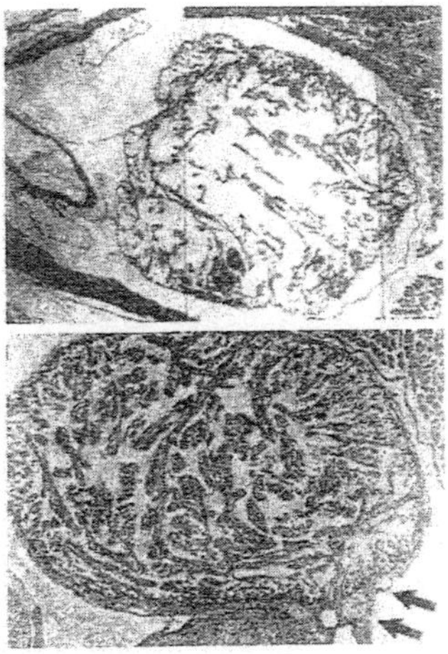

Fig. III-9. The heart of axolotl shortly after landing (upper) and the heart in which the coronary artery is formed in six months after terrestrialization (low, the arrow)

The flow of blood containing richly colloidal nutrients and minerals dissolved entirely speeds up naturally when blood pressure rises. As a result, streaming potential rises. Flow of water alone causes electric potential. Under blood flow containing electrolyte in water, pump-function of the heart alone develops strong electric potential. This electric pulls automatically the trigger of cell genes producing blood vessels or muscles to induce various enzymes. Increase in the volume of oxygen in the blood by 30 times, pulls the trigger of the genes of undifferentiated mesenchymal cells by these enzymes, which induces red blood corpuscles under a certain condition.

Life Evolution Viewed from the Extremities Origin of the Fin

The origin of the limbs is the pectoral and caudal fin of an archaeo-shark. That of Cyclostomata ahead of it is the fold-shaped fin around the branchia. The pectoral fin is for taking food and oxygen, and the caudal fin is for excreting reproductive products of superfluous nutrients and food debris. In an ascidian(mono-segmental animal) further ahead of it, fold of taking-in and excretory pore performing intestinal predation and intestinal respiration are the origin of the fin, and in Pterobranchia it is corresponding to the tentacle. The fin that is originally connected to the gland of the branchia, is an apparatus helping work of the branchia. The function of the branchia is to perform hemopoiesis at the area of the branchial gland (the heart is also a kind of the branchial gland) where oxygen entering into the mouth and nourishing blood returning from the intestine are pumped by the respiratory movement of the branchia. Hemopoiesis is to produce erythrocytes and leucocytes being the blood cell from undifferentiated mesenchymal cells. This is a very important function performed at the epithelial area of the branchia. Another important function is production and excretion of urine being wastes produced by blood function at the mesenchymal tissues (mesoderm) of the branchia, and superfluous nutrients, or reproductive products. This is the urogenital organ among the mesenchymal organs of the branchia, or kidney-adrenal gland-reproductive organ.

The origin of the vertebrate is an ascidian. When a mono-segmental ascidian becomes a multi-segmental chain-salpa type ascidian, the vertebrate quite similar to the fundamental structure of mono-segment appears according to Use and Disuse Theory by cephalad movement. There are a taking-in pore and an excretory pore in an ascidian, and the anus is in contact with the latter.

All branchial glands of Cyclostomata are covered by cartilages. To these cartilages is connected to the cartilages of the fin, and the cartilages of the pectoral fin are connected to the pericardial cavity, and the cartilages of the caudal fin are connected to the pelvic cartilage. Namely, the pectoral fin due to become hands and arms, is in charge of intestinal respiration and the taking-in system of predation, and the caudal fin is in charge of the system of excretion of urine, wastes, and food debris, and reproductive substances. Though as far as the pelvic area is the system of the branchia, the system of the real branchia being the epithelial branchia is as far as the bottom of the pericardial cavity, namely, the diaphragm, and a shark's strap-shaped kidney connects with the bottom of the pericardial cavity, elongating to the caudal side along the dorsal muscle. This is the mesonephros. The kidney of the mammal has been called metanephros, however, it is a lump of the string-shaped kidney that attached to the diaphragm and moved toward the head.

As previously mentioned, thinking of from the view point of clinical phylogeny what the fin (arms and legs) is, you can prevent infants' sudden death perfectly. As long as the limbs or the tongue derived from the branchia are being always moved by any stimulus, babies' sudden death does not happen at all. The reason is that by using a cradle or a teething ring alone, the diaphragmatic nerve (derived from supra-clavicular nerve) via medulla oblongata emits the pulse of rhythmic movement of respiration.

Thinking Of The Origin Of Organs From The View Point Of Clinical Medicine

At the initial stage of creating clinical phylogeny, I have experienced a very impressive case. A professor of a certain university suffered from incurable oral cancer that is too late for treatment, and was referred to me. The tongue cancer became so serious that it progressed to the whole floor of the mouth making it unable to open the mouth. Since operation was no choice to cure, anticancer drugs administration and radiotherapy were performed. By his earnest desire, the patient had received treatment by a qigong practitioner twice in a week in the ward without the hospital's knowledge. As I was known of it secretly, I observed the effects of qigong objectively. As the result of continued qigong even after one treatment course, the cancer had rapidly diminished, and he was able to open his mouth, though radiotherapy and anticancer drugs had been to be ineffective for metastatic focus. He complained of some pain in the focus during projection of Qi. Therefore, I thought that qigong was electromagnetic wave just like body temperature, and took photos of as many as 30 copies. I had observed light having emitted from the qigong practitioner and the patient during Qi projection in about five copies. It has been estimated that life energy's electromagnetic wave worked on the patient's adrenal gland to excrete adrenal cortex hormone by which function cancer cells of metastatic focus were digested by leucocytes. Internal respiration or respiration at the cellular level (internal respiration by mitochondria) is essential for digestion at the cellular level by leucocytes. After I guided him to perform Buddha's diaphragmatic respiration as external respiration, the effect was so great to tell by CT that the cancer diminished. While having kept on performing the respiration lying in a bed, he suddenly came to do just the same exercise as delivery on the way of diaphragmatic respiration, around which time the cancer diminished so greatly that he could tell by himself. He reported; though I am a male, my movements of legs and belly were the same as delivery movement, and my body wanted seemingly to exclude the cancer remaining in the mandible as if to exclude an embryo from the belly.

Since a large cancerous tissue becomes a lump of meat having major histocompatibility antigen just like a mature embryo, continuation of diaphragmatic respiration operates the excretory system of the branchia just like excretion or delivery of a matured embryo consisting of superfluous nutrients, to pull the trigger of delivery movements by employing legs and belly. Just like a male also has mammary gland and breast, he has also genes of linkage-movement of muscles concerning delivery function. I have got an idea of clinical phylogeny by observing this phenomenon. As known by this example, arms and legs are without fail organs of the branchial and excretory intestines in charge of taking-in and excreting in the branchia, being part of respiratory equipments most important for the workings of life.

Chapter Four

Establishment of a New Immunology – Energy Metabolism, Diseases, and Structural Defects of the Human Body

Energy Metabolism And Diseases

Establishing therapeutic medicine of immune diseases while curing diseases

The conventional immunology of self and not-self is based on the histo-immune system developing at the time of transplant, but it does not deal with spontaneously generated diseases. If a new immunology were not therapeutic medicine to cure immune diseases, it would not be worthy of being a science. As Pasteur or Koch had systematized bacteriology and produced the concept of contagious diseases by pathogenic viruses, you have to study diseases well and take measures to elucidate the causes of diseases by a consistent purpose to cure diseases. First of all, considering the mechanisms of life, health, and diseases, let's establish a new system of therapeutic medicine for communicable diseases.

In terms of the mechanisms of life, immune diseases are mainly involved with energy metabolism, or functions. Virchow's cellular pathology had taken the lead in 20th century's medicine. It had been morphology-based medicine to observe the changes of the forms in cells organizing organs or in organs. Morphology is concerned only with materials with mass, but function is mainly concerned with energy, or material with no mass. Until now, no one has researched into diseases in terms of energy. The method of morphology teaches us; if you want to know the essence of an organ, all you have to do is to trace the origin. Immune diseases are of modulated function organs, and modulation of functions can be known by knowing the essence of the organ. Accordingly, all you have to is to trace phylogeny of the organ having lesion.

Originating Clinical Phylogeny

In order to know the mechanisms of diseases, you have to be well versed in phylogeny and ontogeny of all the organs of the human body. Phylogeny means evolution. You have to observe well diseases after elucidation of evolution. By anatomizing diseases for observation and tracing the origin of the focal organ, it is necessary to create clinical phylogeny for knowing the origin of the disease. Elucidation of the evolution of vertebrates leads to finding structural defects existing only in the human being among mammalians. Breathing through the mouth being hindrance to external respiration by acquiring speech, over-gravitational force by erect bipedalism, long –term continuation of estrus by superfluous food intake, the way of eating by swallowing whole without masticating, and wrong food culture of taking cold drinks and foods, are combined to have developed immune diseases.

The author has noticed of strong morphological and functional correlation in various visceral organs between humans and other phyla (vertebrates) observed in ontogeny and phylogeny. And he has applied it to clinical medicine to elucidate the correlation in developing diseases as well as to make use of solving the riddle of evolution, science of which he calls clinical phylogeny. It is a science to elucidate the causes of diseases developing at organs by making use of the following methods; by knowing the origins of human various organs after knowing thoroughly how the origins change by tracing the origins of phyla and life by tracing phylogeny. By this, a clue to establish a new system of immunology has been attained. It may be called the science to research into diseases by Recapitulation theory and the law of Use and Disuse.

Mitochondria (Organelle in Charge of Energy Metabolism) and Diseases

Life is remodelling occurring together with rotation of the vortex of energy, and is the system to overcome aging. Remodelling of the whole of an individual is hereditary phenomenon. No one has studied diseases in terms of energy metabolism and remodelling. Let's think of the relation between energy and diseases.

Development of higher living organisms started at the stage when mitochondria being aerobic prokaryote were parasitic on eukaryotae. Mitochondria are in charge of cellular respiration or energy metabolism. Since the activities of multi-cellular vertebrates among eukaryotae are mostly dependent on energy metabolism of mitochondria being procaryotae, all functio0ns and changes of shapes in higher animals are in the end dependent on the activities of mitochondria. Therefore, health, diseases, toxication and the power of detoxication are all dependent on the functions of mitochondria. The mitochondrial functions are intensely subject to functions of energy with no mass working from outside the body as well as substances with mass taken into the living organisms (nutrients, parasites, poisons, oxygen, gas, minerals and so on). Homeothermal mammals are subject greatly subject to mechanical functions by thermal energy and gravitational one, pressure (atmosphere and water pressure), temperature, sunlight, and radiation.

The medium solution of living organisms is strictly limited to water dissociating electrolytes, and there exist no organisms with oil solvent. This means that essence of life is electric phenomenon (reception or transmission of electron). Therefore, substances with mass are closely influenced by minerals and hydrogen ion concentration. It is in a higher living organism that change in pH of body fluid leads easily to death. Metabolism of hemeothermal mitochondria is easily subject to hindrance by cold energy acting on from outside of the individuals. The essential function of mitochondria is cellular respiration rotating TCA cycle, namely, oxidative phosphorylation (to produce ATP being energy substance). Mitochondria produce important energy essential for life using the membrane of phospholipids. What are essentially required for this function, are Vitamin B, Vitamin C, coenzyme A, and essential fatty acid as well. Acetyl coenzyme A generates mevalonic acid, which polymerizes to generate squalene, which becomes further steroid nucleus within mitochondria to become steroid hormone. The target organ of thirty kinds of cortex hormones as well as adrenalin released from the adrenal medulla is mitochondria. Glycocorticoid (hormone in charge of regulating iron volume and blood glucose) as well as mineral corticoid is essential for rotating TCA cycle. Since mitochondria belong to the same prokaryote as bacteria, antibiotics effective for bacterial microbes have any influence on the function of mitochondria.

The Kidney, Adrenal Gland, and Structural Defects in External Respiration

It is gill-related organs such as the kidney, adrenal gland, pituitary gland, and thyroid gland that act as a mediator between the function of mitochondria (internal respiration) and the function of the branchia and lungs (external respiration). The target of mineral corticoid, Glycocorticoid, and adrenaline in the adrenal gland, is mitochondria. The adrenal gland is the organ to control cell respiration.

Though breathing through the mouth is possible only in human beings, it is the worst structural defect for the human body. When you breathe through the mouth, aerobic resident flora or micoplasma are always taken into leucocytes through M cells of Waldeyer's tonsillar ring, which circulate through the body. And also by cooling the body by cold drinks and foods, enteric bacteria are taken into leucocytes through M cells of intestinal tonsils, which circulate through the body. Since both mitochondria within leucocytes and aerobic bacteria need oxygen, energy metabolism of leucocytes is subject to hindrance.

M cells form vesicles into which they take bacteria; bacteria are digested by leucocytes, and secretory I A (a kind of immunoglobulin) is produced and secreted to saliva or nasal mucus. However, when a lot of bacteria enter with nasal mucus or saliva dry up by breathing through the mouth, complexes brought about by reaction between I A and antigen, are formed and absorbed into the blood, which disturb glomerulus when reaching the kidney by blood circulation. This is immunoglobulin nephropathy caused by oral-respiration (kidney disease necessary for dialysis called nephrosis).

Since the kidney, adrenal gland, and organs of the urogenital system, are all originally equipment for the mesenchymal system of the branchia, the urerus, urinary bladder, and prostate gland, as well as the organs of the kidney and adrenal gland system, suffer often damage. It is a group of diseases called autoimmune diseases conventionally that occur with latent infection of leucocytes or histocytes by resident flora, virus, or mycoplasma. As a result of having elucidated the mechanism of the kidney and adrenal gland connecting internal respiration with external respiration, I could establish as a matter of course a new system of immunology.

Exploring Energy Factors In Diseases Via Therapy

The author has clarified that evolution of vertebrates has occurred according to the law of Use and Disuse under the leadership of biomechanics based on gravitational function, by having developed artificial bone marrow chambers and artificial dental roots. Once it has been clarified that evolution has occurred under gravitational energy according to Lamarck's Use and Disuse law, it is easily understood that your body condition becomes good or bad depending on energy, you get ill under inappropriate energy. It is necessary to observe well the relationship between structural defects of the body peculiar to humans and modulation in energy metabolism. In order to establish a new immune-therapy, you have to observe diseases carefully from the viewpoint of energy in addition to conventional aetiology, and remove energy functioning inappropriately on the body. I want to show the cases by which I could have established clinical phylogeny, and the cases in which treatments were effective under the application of this study.

Examples (Case 1)

A male of 73 years old. Systemic itching exanthema and swelling of gums. During hospitalizing about one year at the department of internal medicine, Tokyo university hospital under the diagnosis of leukaemia and dementia, he was referred to the department of oral surgery. At the time of consultation, he suffered walking difficulty, and could not hold a conversation. His gums had swollen as a whole and formed swelling tumorously. He was very anxious for extraction of teeth and could understand what was spoken, which testified that he did not suffer dementia. After deliberate extraction, gum's swelling disappeared, followed by disappearance of exanthema. He came to hold a conversation, and leukaemia was healed completely, resulting in discharging with delight. In the department of internal medicine, the diagnosis was changed from leukaemia to osteomyelodysplasia.

Let's think this case out. Severe gingival infection spreads into the bone marrow of the jawbone supporting teeth. Though normal elderly marrow cavity of the jaw bone is fatty and is not in charge of haemopoiesis, strong inflammation induces leukocyte haemopoiesis, with which leucocytes a large amount of saprophytic bacteria and bacteria causing periodontal disease are infected. Since at the process of ontogeny the jawbone is divided from the dermal bone, which is derived from the ectodermal skin, infected leucocytes disorder the skin, brain, eyes, nose, and ears that are organs derived from the ectoderm. That's why exanthema, dementia, and leucocytes filled with bacteria misdiagnosed as leukaemia in the internal medicine) were cured through extraction alone. You will be able to understand by this that if you reckless things with atopic dermatitis, inflammation will occur in the retina resulting in losing your sight.

(Case 2)

A seventy years old female living in Chiyoda ward. She fainted at the time of scratching itchy places by sudden histicaria throughout the body when she was waiting for a bus, and was taken to the hospital in an ambulance. By and by, she often suffered urticaria and faint. However, urticaria did not develop once for all since she changed into nasal respiration after having attended a lecture by the author at the centre for the elderly in the same ward that oral respiration is the origin of every disease. And she visited me and consulted the department of oral surgery searching for right living ways. Correction of nasal respiration, respiratory exercise, and sleeping posture made her high-spirited. If you inhale exhaust gas through your nose, it will be purified, not causing urticaria. In this case, exhaust gas was inhaled through the mouth, and taken directly into leucocytes. After three minutes after respiration, it reached the subcutaneous tissue, where maldigestion's reaction occurred to produce histamine. As the result, urticaria developed. The skin is the system to excrete toxins or superfluous chlorides since the age of primordial vertebrates. Since the brain and nerve are derived from the ectoderm, urticaria in the region of the interbrain glia corresponding to the subcutaneous tissue develops faint. The doctor unaware of clinical phylogeny has made a fuss over this disease, having called this disease a strange one.

(Case 3)

A four years old boy who had been hospitalized in paediatrics department with symptoms such as myasthenia gravis or myodystrophy were referred to me for treatment of stomatitis and caries. Having proceeded with oral treatment improving bad oral respiration to better respiration, myodystrophy was greatly improved. According to his mother's detailed comments on his medical record, meconium (faeces produced during viviparity, which are defecated after delivery) entered the lungs to cause a strange disease such as this. Since patients with myasthenia gravis perform oral respiration by 100 percent, their lymphatic rings of Waldeyer are infected with indigenous aerobic bacteria. When these infected leucocytes contaminate the thymus gland, this disease is considered to develop. Since Waldeyer's lymphatic ring derives from the second branchial gland, and the thymus from the fourth branchial gland, this disease has been brought about by spread of infection in the profound associations among organs of the same origin.

This case shows that the lungs contaminated by meconium take on the similar symptom as myasthenia gravis with the thymus infected alone; since the lungs are derived from the sixth branchial gland, the thymus derived from the fourth gland is naturally contaminated as well.

(Case 4)

A male of 45 years old previously mentioned who had lost his sight 13 years ago because of unknown causes, consulted me after hearing "Health is determined by respiration" that was transformed into voiced version. As I found atopic dermatitis in his face, I asked him whether he took harsh exercise or not; he played American football and Rugby in his school days, and three years later after having become a working man he was aware of having lost his sight, because he could not see traffic signals suddenly. By converting oral respiration to nasal one, taking a good rest, and masticating well, he came to be able to tell the shadow of a person apart faintly, and recovered to the extent to discriminate colours.

The nerves in the region of the brain, eyes, nose and ears, and the skin and hairs, are derived from the ectoderm. Accordingly, in this case atopy has just occurred in the retina. In sports, overweight gravity functions on skeletons. Since man stands upright, he is more likely to be susceptible to larger gravity than the dog or cat. As a result, haemopoiesis in the bone marrow is disturbed. Combination of oral respiration, cooling the gut by taking cold drinks, and lack of taking rest, develops easily atopy, which furthermore develops serious immune diseases. It is because overweight gravity weakens leukocyte-haemopoiesis.

(Case 5)

A female of 65 years old with diabetes and periodontal disease consulted me. Improvement of worse oral respiration, treatment of periodontal disease, and irradiation of artificial solar light to the region of the sole and head (30 minutes), made diabetes disappear completely.

When periodontopathic bacteria and aerobic bacteria by oral

respiration are carried hematogeneously to the pancreas by leucocytes, diabetes develops. Solar light excitates haemoglobin and cytochrome (respiratory protein composed of iron and protein, which transmit electrons) to activate mitochondria, elevate immune capability, and reinforce the digestive power of leucocytes.

(Case 6)

A five years old boy with epilepsy and asthma consulted me. Since asthma was caused only by oral respiration, there happened any seizure for more than 7 months after employment of a teething ring. Epileptic stroke was pointed out by the abnormality in EEG at the paediatric department, and one year has passed since taking antiapasm0odic. Inquiry revealed that light anaphylaxis fit occurred by proteins such as raw fish or shells, because he was given raw shrimps since one and half years of age. Taking care not to give shrimps, shells, dried fish and meat, artificial solar light was irradiated, resulting in no fit at all.

Proteins should not be given to sucklings of mammals. Proteins given are absorbed intact to become antigens, producing antibodies at an appropriate time to be responsible for allergy or anaphylaxis. Since especially raw shrimps, shells, raw fish, peanut butter and brown needle made from buckwheat flour may cause anaphylaxis to death, you have to take notice not to give them at all until two and half years of age. These proteins are probably used in the mitochondria of infants7 cells, and activate the function of mitochondria at first. When antibodies are produced one or two years later, the electron transport system of mitochondria is disturbed, and convulsion occurs once any shrimps being antigens are given. The abnormality of mitochondria leads to abnormal EEG. If you know only the characteristics of the intestinal membrane of mammalian sucklings, food-allergy in infants will not be completely avoided.

(Case 7)

A female of twenty seven years old. She suffered palindromic rheumatic disease, colitis, and headache. She had bad oral respiration with a blush on her face. Improvement of respiration, selection of food, well mastication, giving up cold drinks and food, taking a good rest, rendered all the symptoms disappear, resulting in full recovery.

Taking cold drinks together with oral respiration causes oxygen deficiency in the portal blood of the intestine to develop headache or igraine. When further advanced, depression develops. Oral respiration hurts the skin, cartilage, intestine, lungs, pancreas, and uterus by contaminated leucocytes with resident flora and so on. If overweight gravity functioning on the body, and inappropriate thermal energy, are controlled systematically, immune diseases are easily prevented.

(Case 8)

A female of twenty-four years old. She was referred to me under the disease of ulcerative colitis. Improvement of oral respiration, refraining from cold drinks and foods, intake of easily digestive food, masticating 30 times at one bite, sleeping for nine hours to compensate for lack of rest, made her restore completely after three months.

(Case 9)

A thirteen years old girl. She suffered atopic dermatitis since seven months after birth. At present she is dark-skinned, with the inside of her elbow and knee having eruption. She is so short that she passes as a third grade student of an elementary school. Sideway sleeping posture, unilateral biting, and oral respiration were corrected. And changing sleeping hours from seven hours to nine ones, implement of respiration exercise and gargling often, and cleaning her skin with chlorides, made her skin white, not-dark-skinned, and eruption disappeared. During a winter vacation, she was forced to sleep 13 hours for two weeks. As a result, her height grew by 6 centimetres and her skin became white to her great joy.

(Case 10)

A female of 67 years old. I lectured on “Health is determined by respiration” at Gakushikaikann hall dinner party. An elder sister of president of a certain university who read the bulletin consulted me hoping to treat the lungs buy changing oral respiration to nasal one, because she was told at a public hospital that her illness named bronchiectasis would not improve further, but asked to keep on taking medicine. I guided her to use a nose-lift, a breath-trainer (a teething ring for an adult), and a labial tape during sleep, to lie on her back using spongy pillow, and gave her phototherapy. After one week, coughs and sputum reduced by half. As conspicuous improvement was obtained one month later, taking medicine of Claryth (antibiotics effective for micoplasma in the lungs) was reduced. After that the symptoms was improved without taking medicine.

(Case 11)

A vocalist of 62 years old. He had been treated for interstitial pneumonia at a famous university's medical hospital. Since keeping cold drinks and food off, training of mastication, respiratory exercise, correction of sleeping posture, and artificial sun beam phototherapy as well as guidance of nasal respiration and of food contents, had brought about great improvement, I guided her to reduce the amount of drugs administered by the internal medicine. As a result, the condition of the disease improved further, and both coughs and sputum disappeared. When the course of the disease had been well, he consulted the internist. He was asked, "Are you taking medicine now?" When he answered "I am giving up taking medicine now", the doctor changed his countenance. "If you don't take medicine, I will not consult you any more", was he told. And looking into the clinical data, the doctor said, "Since the clinical data are getting well, today I'll consult you, but If you don't, I will never consult you at all". He was quite at a loss, and consulted me how to do. I told him, "Don't take medicine, preserve the drugs when necessary, and tell the doctor that I am taking medicine". Every time he told the doctor the same saying, the doctor was in a very good mood, sayings "the drug is getting very effective". At the time of a sequence overwork, and recurrence of cough and sputum, I advised him to take the preserved drug for short periods and take rest. He recovered soon. Later, I implanted nine artificial dental roots at a time to restore his masticatory function, which healed him of pneumonia further.

Both the case 10 and 11 are mycoplasma pneumonia. Mycoplasma is an aerobic parasite in the respiratory tract. If it parasitizes the cell of leucocytes or tissues to activate, it competes in oxygen essential to oxidative phosphorylation in mitochondria, disturbing cell respiration. For this treatment, prevention from infection at the tonsillar lymphatic ring through nasal respiration, and activation of mitochondria by phototherapy, are very effective.

(Case 12)

A twenty-six years old female. She has been taken often care of ambulance with panic reaction caused by hyperpnea, and had the habit of oral respiration, unilateral mastication, and lying on her side during sleep. Strict guidance on nasal respiration exercise as well as correcting these habits, refraining from cold drinks and food, and taking drinks and food of 40 centigrade, did not cause onset of anxiety neurosis and hyperpnea at all. At the stage of complete recovery, she came along with her two daughters (7 and 5 years old), hoping the treatment of their oral respiration. I made them to use a teething ring, which led easily to cure. At the next consulting time, she reported with pleasure crying: In fact I was so much bothered by child rearing that I suffered panic reaction and became an alcoholic, but I recovered even alcohol poisoning according to the author's guidance.

In panic syndrome, chest respiration is performed by will based on the sympathetic nerve and light respiration by the mouth not by the nose continues differing from the diaphragmatic respiration to activate slowly the body by the primary parasympathetic nerve. This is easily cured by guiding nasal respiration and diaphragmatic respiration, performing mild respiratory exercise, and warming the gut. This disease as well as depression or migraine is caused by oral respiration and deficiency of oxygen in the internal organs such as the gut. As alcohol poisoning is also caused by the deficiency of oxygen in the internal organs with cold drinks, these two troublesome diseases can be cured completely by the methods to perform nasal and diaphragmatic respiration, and warm the gut.

Aerobic bacteria taken into leucocytes from M cells the Waldeyer's tonsillar ring, circulated through the body; If the large intestine becomes infected by them, colitis occurs; If the pancreas, diabetes, If the hemopoietic nests of the leucocytes, arthritis or rheumatism. This is because mitochondria were originally aerobic prokaryote settled in eucaryote in ancient times, and aerobic bacteria ate oxygen of mitochondria so that energy-metabolism was disturbed and at a result bacteria could not be digested or sterilized within leucocytes. Especially if the body temperature is lowered by 1C, mitochondria within the cells of the vertebrate stop almost energy-metabolism. That's why cooling the gut leads to upset stomach. If you continue to cool the gut since the time of a baby, you suffer chronic indigestion. Thus in today, low temperature, disturbance in growth, and bad physical condition give birth to cold-blooded fellows.

As previously stated, thermal energy influencing directly the functions of mitochondria and over-gravitational function (lack of bone rest) as well as latent infection by indigenous bacteria with oral respiration are deeply involved in occurring immune diseases. It is above 12 cases that have been treated by controlling all these causes.

To summarize immune diseases, many of communicable diseases referred to as classic diseases do not produce inborn immunity, so that vaccines and the like have been made with various ideas. Occurrence of infectious diseases means coexistence of higher animals and micro-organisms. Enteric bacteria can be also easily absorbed into the blood. Too many of them cause trouble. Therefore, immune capability refers to the power of remodelling, or the power of metabolism at the cellular level. I stated at the beginning part of this

Chapter that the organs intervening the lungs in external respiration and mitochondria in internal respiration were uro-reproductive organs such as the kidney and adrenal gland. The intervention has been mainly carried out by steroid hormones. This fact has been unknown. In immune diseases, the body as a whole is wrong with health due to unknown reasons, accompanying boredom, feebleness, slight fever or shortness of breath. This is caused by the disturbance in energy metabolism at cell organelle, mitochondria. It has been empirically known that steroid hormones have been effective for these symptoms. That's why steroid hormones have been use recklessly for immune diseases.

The Mechanism of Diseases

Diseases based on conventional pathology have been classified into abnormality, trauma, infectious diseases, inflammation, cysts, tumour, and the others. This is a classification due to morphology mainly based on cellular pathology. Immune diseases have been odd and strange diseases of unknown origin among the others. Considering that immune diseases are derived from disturbance in energy metabolism, that energy poured into the body is the cause of diseases, and that infection by indigenous bacteria is a factor to cause diseases by disturbing energy mechanism of the body, I have cured the previously menti9oned cases of unknown origin by controlling energy and infection.

Originally, "Immune" meant to protect against an epidemic. It has already been known that it owes roughly to the function of blood and the cause of diseases is derived from bacteria, when it was clarified that the substances to oppose these diseases are produced in serum among blood. Even if one is infected with an epidemic, one can be cured to be alive without dying. Then, immune substance can be produced in serum, which is immune serum. It was Edward Jenner who discovered that immune capability for small pox is produced, when cowpox vaccine that is not so much harmful to a man as well as a cow is inoculated to the human body. After that, there were times when treatment or prevention was carried out using diphtheria immune serum produced by inoculating diphtheria bacillus to a horse. However, since antibody is produced in the serum of a horse, it was proven that the same person could not use horse-antiserum many times.

As medicine has been systematized, it has proven that there are various kinds of diseases except communicable diseases. The concept of immune capability has shifted to protecting the body as well as curing diseases. Diseases have the following seven kinds. 1. infectious disease 2. dietary deficiency 3. disturbance in energy-metabolism 4. disturbance in remodelling and differentiation (cancer and deformity) 5. function of poison 6. incompatible organ transplant 7. trauma

The causes of these diseases roughly consist of the following seven kinds.

1. Infectious diseases: infection by pathogenic bacteria, virus, protozoans, and parasites, and latent infection by common enteromicrobes or non-pathogenic bacteria.
2. Malnutrition: deficiency or oversupply of substances with mass such as nutrients, vitamins, minerals, or oxygen.
3. Failure in energy matabolism.
4. Overwork and inappropriate intake of energy: gravitation, temperature, atmospheric pressure, humidity, sun-light, radiation, electromagnetic wave, a sound wave, an ultrasonic wave.
5. Function by toxic substance: poisons, exhaust gas, toxins, or agricultural chemicals.
6. Histogeneos immunity reaction accompanied by transplant of internal organs, organs, or tissues.
7. Injury: external force.

This science had been called serology before called immunology, at which times certain diseases had been treated by antiserum. Viewed from the classification of diseases, immune capability in those times turned out to have been effective only for pathogenic bacteria among infectious diseases. Today's self-notself immunology started at chimera by a quail and a chicken at the embryonic stage by Le Douarin, which naturally raised only questions about histogenous immunity. Immune diseases and immunology can not be discussed without including from allergy to syncope accompanied with hives, dementia concurrent with dermal eruption, death from overwork, and chronic fatigue among contemporary diseases. The reason is that people suffering from immune diseases are without fail exhausted. Immune capability or the power not to fall ill, refers to one's life force at the cellular level. Metabolism force by which an individual develops after fertilization, differentiates, grows and remodels, is one's force at the cellular level, or one's defence force against illness. The growth force and remodelling power of cells occur together with vortex (metabolism) of energy. Almost vortex of energy is supplied by cellular respiration or internal respiration. In order to know immune capability, it is necessary to know cellular respiration. When this immune capability is disturbed by previously mentioned

1. Infectious diseases
2. Malnutrition
3. Failure in energy production
4. Overwork
5. Toxins

Immune diseases develop. Atopic dermatitis, asthma, sarcoidosis, bronchitis, cardiomyopathy, or arteriosclerosis occurs by one or a combination of the five kinds, almost cases of which have latent infection by indigenous bacteria at the base. Accordingly, immunology can be referred to as the system of medicine: to elucidate the causes of diseases occurred already, and to remove them to cure as well as to control the five kinds to prevent diseases. Diagnosis of the causes and treatment of diseases are two sides of the same coin. That is, therapeutic diagnosis clarifies causes.

New Immunology

Life phenomenon is the system to conquer the aging of parts of an individual by remodelling occurring together with rotation of energy vortex by burning nutrients with oxygen in the function of time, within colloid (corpuscle as large as a molecule) of substance having mass in the phase of solid, liquid, and vapour in aqua that is bounded by semi permeable membrane of phospholipids. Remodelling of an individual as a whole is hereditary phenomena, which is carried out through reproduction. Solvent of life is limited to water that is the standard of all substances (specific gravity 1, specific heat 1, ice point 0 °C, boiling point 100 °C). There is no life of which solvent is oil. Water has the ability to dissociate electrolytes, and oil can't dissociate them. That is, life phenomenon is electric phenomena within colloid. In multicellular organisms, the epithelial and nerve system, and the hormon system are controlled by the membrane potential at the ionic channel , and by streaming potential are controlled the muscular, skeletal, and cardiovascular systems belonging to the mesenchymal system. In the complex constitution of vertebrates, membrane potential, hormone, and bodily mechanical movements are converted into streaming potential through fluid dynamics, which pull the trigger of cell genes to cause cell differentiation. By this, cells having the same sity trillion genes form various organs or tissues having innumerable kinds of shape and functions, to make an integrated individual as a whole. Roughly, three kinds of life exists on the earth: prokaryotes, eukaryotes, and multicellular organisms. I have already mentioned this very often. Since in conventional life science, these fundamental points have been ignored so much, and have been unfamiliar, I would like you to keep them in mind.

In order to rotate energy vortex, life takes high-energy substance (nutrients) from life media into the body, and use the power produced by dissolving it. Vortex of life-force is subject to energy functioning outside life. Procaryotes and eukaryotes are scarcely influenced by gravitational force or thermal energy (between 4°C – 40°C), while multi-cellular living organisms are strongly influenced. The method to take nutrients into one piece of cell differs among three kinds. Procaryotes absorb directly nutrients within medium to rotate the vortex of life force. Eukaryotes absorb by taking nutrients as food vacuole (vacuole surrounding food taken into cells) as well as directly absorb nutrients within medium. Directly absorbed substances are consumed by small organelle, mitochondria to rotate respiratory metabolism. Mitochondria are aerobic prokaryotes parasited on eukaryotes, which perform respiration or perform oxidative phosphorylation to rotate TCA cycle. IN multi-cellular living organisms, eaten foods are absorbed in the intestines, and the blood carries these nutrients to each cell, which absorbs directly usable nutrients from the blood, which mitochondria use. Moreover, they absorb nutrients by pinocytosis to re-digest within cells and consume. Multi-cellular organisms re-digest at the cellular level what the blood has brought, and provide them for remodelling as well as cell respiration. This power is the life force of cells, the power to conquer diseases (free from illness), that is , immune capability.

Obstacles to mitochondria in cellular respiration lead directly to disturbance of formation of collagen, cartilage and bone, but also develop disturbance of cell-remodelling. This is the immune disease. Complete deficiency of Vitamin B_1 stops the function of mitochondria to cause instant death. This is the so-called beriberi heart. Since the vertebrate has characteristically osseous spine and intestinal respiratory apparatus, if you elucidate the relation between intestinal respiration and cell respiration, and the enigma of the development of bone marrow hemopoiesis, a riddle of the vertebrate will be solved all at once.

The vehicle of oxygen in cell respiration is an erythrocyte, which originally develops at the intestines. If you clarify the causal factors why this moves to bone marrow of the skeletal organ, you will solve a riddle of the immune system.

The author, as he mentioned the causal factors in the preface, has clarified that they owe to elevated streaming potential (1.5-10 microampere) accompanied by rise of blood pressure as a result of response to gravitational energy by having developed an artificial bone marrow equipment through model study. As a result, I have elucidated for the first time in the world that immune diseases were caused by lack of rest by long-standing posture (over-gravitational force), by mal-intake of energy by material with no mass such as cooling the gut by cold sake or wine, and by latent infection with saprophytic bacteria by failure in external respiration being the pivot of energy metabolism, and moreover disturbance in the metabolism of mitochondria being the pivot of cell respiration were caused by injury in the nutrient-absorption system in the intestine and latent infection by indigenous bacteria in the intestines or the throat.

By this, the enigma of the development of infant's baby food allergy (infant's atopic dermatitis by protein having antigenicity) as well as that of the development of atopy by mothers' milk, has been clarified. On the other hand, in the case of serious diabetes representative of metabolic disease, mitochondria deterioriate. That is to say, mitochondria have controlled all of the body ranging from the metabolism of glucose, lipid or protein to sex hormone, and uroexcretory system. Modulation of mitochondria has proven to be the essence of every immune disease. Thus, the riddle of the development of allergy and that of the development of immune disease caused by disturbance in remodelling has been elucidated. The foundation of metabolism in multi-cellular organisms lies in eukaryotic protozoan cells, the source of which metabolism lies in the respiratory system of prokaryotes, or energy metabolism of mitochondria. In other words, immune diseases have been caused by the disturbance of respiratory function in mitochondria with inappropriate gravity, temperature, and light-energy functioning from the outside, and parasites coming into the blood through the intestinal membrane (bacteria, virus, protozoa, and parasites), and with protein, hormone, amines, antibiotics, toxin (exhaust gas) and so on. As a result, growth, development, and remodelling at the cellular level have been disturbed, followed by the break down of nerve cell, muscular tissue, subcutaneous tissue, cartilage cells, blood vessels, or various kinds of organs.

Mitochondria with the Internal Respiration System and the Branchial Intestinal Muscular System with External Respiration

The protozoan being the uni-cellular organism is the eucaryote, which has generally a cell organelle (mitochondria) within a cell. The system of external respiration of the protozoan draws on the surface of cells, while internal respiration draws on energy-generation reaction in which oxygen and glucose taken into from the cell membrane rotate TCA cycle by oxidative phosphorylation at mitochondria. Almost all life activities of the animal are provided by energy generated by mitochondria. The protozoan has no system to intervene between external respiration and internal one, because it absorbs at the cell membrane directly nutrients and oxygen necessary to mitochondria through media.

Animals except the vertebrates perform dermal respiration similar to that of the protozoan; they absorb oxygen from the skin, and carry the nutrients absorbed from the intestine into the cells of the whole body through blood of the cardio circulatory system attached to the intestine, which are absorbed to produce energy-substance (ATP) at mitochondria; that is, cellular respiration is performed. On the other hand, only the vertebrate performs intestinal respiration (from the branchial intestine to the lung), and has an osseous spinal column. Change of the form in the animal evolution is subject to Wolff's law based on biomechanics, once habitual ways of moving are determined. The vertebrate separated from the others since Pterobranchia (Bryozoa, Bugula) had induced the branchial respiratory epithelium at the ala into the intestinal duct by the way of intestinal duct predation. The source of intestinal respiratory movement is repetitive exercise of the wave. In the vertebrate only, part of the movement of the branchial gland in branchial respiration becomes the circulatory system. Namely, in the vertebrate only, in the system of external respiration is, the heart included.

The branchia of a hagfish (Cyclostomata) moves amorphously just like the heart. Down to the shark of the primordial vertebrate, one of jawed vertebrates, its branchial heart united, and each branchial intestinal respiratory visceral muscle adhered to the branchial arch bone and united to form the immovable tongue. The heart existed at every branchi in the centre of the tongue retracted to form the branchial artery, in the base of which there is the heart surrounded by the pericardial cavity. At cyclostome, every branchi is covered by thin cartilage which is taken over by cartilage surrounding the branchial arch and the heart of the shark. The cartilage (clavicle) connecting to the fin also forms the abdominal side of the pericardial cavity, and the heart is designed to move by moving the fin, because both the fin and clavicle were originally the branchial arch. That's the reason why the arms ache when the heart worsens.

Though it has been conventionally said that the respiratory visceral smooth muscles of the vertebrate had become masticatory muscles, swallowing ones, vocal ones, auditory ones, facial expression ones, and have no muscles peculiar to respiration, the region from the tongue muscle to the caudal bottom of the pericardial cavity is the respiratory visceral muscles. Accordingly, the thoracic fin destined to become the arm, the anal fin to become the leg, and the diaphragm, are part of the branchial respiratory system. The lungs in the reptile are formed to the pelvic region avoiding the pericardial cavity, and respiratory exercise are performed mainly using the muscles of the pharyngeal muscles and abdominal wall, which were originally the branchial arch muscle.

Mechanism Of The Body And Health System Connecting Internal Respiration to External One

The mono-cellular protozoan differs from the multi-cellular vertebrate in the bodily mechanism. Differing from the protozoan in which organelles having various kinds of function within the cell are united, the latter is composed by combining various organs composed of cell groups of which functions are differentiated. It is the vertebrate that is the chordate having the osseous spine and performs intestinal respiration. Bone and intestinal respiration are characteristic of the vertebrate; the prototype of intestinal respiration is the branchia, and that of bone is cartilage. The branchia is a haemopoietic apparatus performing erythrocyte-haemopoesis at the epitherium and the mesodermal system, and cartilage is the product of energy-metabolism at mitochondria doing cell respiration and the material as well.

Since from the endodermal intestinal epithelium and ectodermal epithelium, skin or respiratory membrane, are unnecessary products by metabolism under the epithelium excreted, the branchia also performs excretion as well as takes in oxygen. The excretory system of the intestine and parietal muscles is the kidney-adrenal gland system existing in the branchia of the chordate being the primordial vertebrate.

Since this controls the metabolism including energy of the whole body and its products as well as excretion of the metabolic products of the muscles. In the aquatic age, it excreted carbon dioxide solved in water, and what's more controls reproductive substance being superfluous nutrients, fats, and minerals. This is the urogenital organ, and was in fact part of the system of the branchia. Since the branchia is in charge of respiration, erythropoiesis is performed at the branchia under energy production by active mitochondria in the primordial vertebrate. Since mitochondria excrete cartilage at the same time of energy metabolism, every branchial gland of cyclostome is covered by cartilage. Collagen and cartilage are metabolites of mitochondria and are substances for metabolic turnover. Haemopoesis in the mesoderm of the branchial intestine and the kidney-adrenal gland-urogenital system in charge of storage and excretion of products brought about by assimilation, metabolism, and catabolism of the muscles, are the system connecting external respiration with internal respiration. Namely, it is the equipment to support respiration at the cellular level of the whole body that constitutes part of haemopoietic apparatus being the essence of the branchia. As the remnant, in the human kidney is produced cytokine (enzyme protein) inducing erythropoiesis called erythropoietin. The kidney filtrates to excrete unnecessary minerals or materials carried by blood from the muscle-skeletal system and every organ of the body.

The adrenal gland is divided into the medulla and cortex with each having different function. From the medulla, adrenalin and noradrenalin having sympathetic nerve function acceralating metabolism of glucose and lipid through cardio-vascular system as the endocrine hormone are secreted, while from the cortex glucocorticoid, mineralcorticoid, and sex hormone (androgen) are secreted into the blood. Glucocorticoid controls every intermediary metabolism of three typical nutrients of glucose, lipid, and protein, while mineralcorticoid controls excretion and absorption of Na^+, H^+, Cl^+, K^+ by acting on the kidney. It is the sex hormone, androgen that stimulates the development of the apparatus or organ to control the induction of differentiation of reproductive cells and superfluous nutrients (lipid) made by glucocorticoid. As stated at the beginning, life is a colloidal system to remodel by receiving and transferring ions through rotating vortex of energy by burning three representative nutrients. Glucocorticoid is in charge of energy, and mineralcorticoid performs metabolism of nutrient salts (K^+, Na^+, Mg^+), Ca^+, and H^+ within mitochondria to control. The hormone in charge of the essence of life phenomena is parasympathomimetric (cholinergic) adrenal gland cortex hormone, and it is sympathomimetic medullary hormon that activates the body by raising blood pressure through raising blood concentration of glucose and lipids. It is the previously mentioned mitochondria that produce ATP by rotating Krebs cycle using nutrient salts and oxygen by means of cytochrome respiratory enzyme by dissolving three representative nutrients including glucose into pyruvic acid.

Therefore, the adrenal gland is the generator of energy in which oxygen and minerals taken from the branchia (external respiration) by adrenal gland's hormone and nutrients absorbed from the intestines actually function. Every metabolism being the source of the vertebrate is almost 100 % dependent on mitochondria and the work of adrenal hormone to control their function. Superfluous energy produced thus is stored as superfluous nutrients of which part becomes reproductive cells as well as consumed in cell-remodelling, namely, in metabolism. Every conversion is controlled by more than 30 kinds of steroid hormones. Steroid hormones are produced as follows: mevalonic acid is synthesized from coenzyme A within mitochondria, which is condensed to become by turns squalene, cholesterol, steroid nucleus, and finally steroid hormone. Adrenalin, medullary hormone secreted by the sympathetic nerve controls the prior to the stage in which glucose is metabolized by mitochondria. That's why the primordial shark whose sympathetic nerve does not develop yet, has naturally no adrenal medulla. Since the essence of life phenomena, as often stated, is "remodelling occurring with rotation of energy vortex, the kidney performs anabolism, metabolism, and catabolism of nutrients absorbed from the intestine through the blood, and transforms them into urine, carbon dioxide, fat, and reproductive cells may be referred to as the organ implying the real meaning of life.

Essential for cell respiration of mitochondria are Vitamin B_1, coenzyme A, NAD (nicotinamide adrenaline dinucleotide, which contain nucleic acid-base (pyrimidine, adenine), sulphur, and phosphoric acid. Acetylcholine is also produced by acetyl CoA. When respiration of mitochondria rotates, cartilages including much sulphur as metabolites are excreted, and when respiration jumps, metabolism of minerals is activated, and calcium triphosphate is added to cartilage, which crystallizes in vivo to become apatite. Cartilaginous dermal denticle develops at the skin where cellular respiration of an ascidian is most prospering. By this reason, the surrounding of the branchia of Cyclostomata is completely covered by cartilage. These cartilages become apatite by leap of cellular respiration accompanied by activated activity of the vertebrate and ossifies, which is also due to acceleration of mitochondrial function.

Health Is Determined By Energy Metabolism

Understanding the constitution of the vertebrate well, you recognize easily that in the vertebrate, including human beings, the branchial intestine absorbing nutrients and oxygen extended to the pelvic region. It is the erythrocyte and leukocyte that mediate between the lungs, transformed external respiratory branchia, and cellular respiratory mitochondria existing in each cell in the range of 60 trillion in adults. It is adrenal cortex hormone that directly controls the function of mitochondria, and it is adrenal medulla hormone that that controls indirectly by blood pressure or blood sugar. Namely, the target organ by the hormones secreted from the uro-reproductive system of the mesenchymal organs in the branchia of the external respiratory system have been 100% mitochondria in cellular respiration. And the organ connecting osseous vertebral column that is characteristic of the vertebrate with the branchia of intestinal respiratory organ (the lungs) has been mitochondria with cellular respiration.

Since bone has phosphoric acid, calcium, and nutrient chlorides essential for mitochondrian function, mitochondria perform oxidative phosphorylation by obtaining materials from bone, excrete their products as cartilage or bone, and are in conjugate with bone throughout life. That's the reason why neglecting bone-rest breaks down the bone marrow haemopoietic system, and as a result respiratory cells are inhibited.

It is because energy production and metabolism by smooth cell respiration are necessary for every cell life activity.

Since the greatest achievement in the 20th century's science, the 'law of energy conservation' has not survived even in today, contemporary medicine has been established under "the law of conservation of mass" of the 19th century. This is the American medicine on an organ by organ basis derived from Viorchow's bacteriopathology in which morphology is centred; Viewed from immune science curing illness, it has been self and not-self immunology quite different from immune diseases. I've showed here the achievement obtained by the study of health and illness from the viewpoint of energy metabolism based on the law of energy conservation. Viewed from energy metabolism and massless substance called energy, a human individual is a living existence of complete open system adhering to the earth surface floating in outer space. Due to the misunderstanding that living organisms have been produced only by substance having mass, the human body was mistakenly recognized as a semi-closed system, which rendered 20th century medicine to a standstill. That's why immune diseases have not been cured. Conquering immune diseases can be attained at last by controlling energy metabolism.

Structural Defects of the Human Body A New Concept of Life

Having reorganized medicine in terms of energy is the establishment of new immunology. Medicine based on "the law of conservation of energy" has been recognized at length in the 21st century, breaking away from "law of conservation of mass" in the 19th century. The author has obtained "What is life?" through a new outlook on the universe, and established the system of medicine as to a new way of living; the author has filled a leather bag with new sake (rice wine). It is not so simple to obtain the law of conservation of energy. As long as guts are alive perfectly, mind being essence of life is alive. Once you understand that mind is life energy which is substance with no mass being essential for life, brain death would be a wrong thought, viewed from the right outlook on life. According to the high ethical thinking of life based on the law of conservation of energy, we can not help denying organ transplants. If you are aware that energy is substance with no mass, evolution of life will be elucidated so that you'll find out whereabouts of the heart. Then, you'll understand what has made this world of the 21st century unstable. Medicine has advanced to organ transplants today. Though death of the body has to be consistent with death of heart or soul, it has split today.

Now that it has been clarified that evolution occurs by energy, detailed investigation of the human body has showed five structural defects as roughly mentioned previously.

Chapter: 1. To be able to breathe by the mouth causes latent infection, which leads to immune diseases. 2. excessive gravitational function by erect bipedalism causes deficiency of bone-rest, which inhibits remodelling of the haemopoietic system to inhibit cell respiration. 3. Sufficient food clothing, and shelter as a result of accumulation for 6 million years, and the long term mating season by superfluous nutrients, have caused turbulence in reproductive behaviour in the civilized country. 4. Medical advance in the wrong direction by wrong outlook on life. 5. Over-pursuit of pleasant sensations by development of the cerebrum in the advanced countries. In the following paragraph, I'll comment on errors in medicine and over-development of the cerebrum.

Medical Care by Wrong Outlook on Life. Errors in Nursing and Rearing

What has to be questioned in our country is errors in nursing. The author describes how children raised by this error develop through childhood, and adolescence. When you give baby food at the fifth to sixth month after birth, or give cold milk, babies are obliged to perform oral respiration and suffer green stool (baby's stool with cement-like colour by maldigestion. Baby's stool should be gold in colour.). Continuous green stool for one or two years leads to hypothermia in the childhood, causing disorder of the intestine through atopic dermatitis, and asthma. And in severe cases, ulcerative colitis, or Crohn's disease (terminal ileitis) will develop around 20 years of age. Cooling the intestine paralyzes the sympathetic nerve and makes emotion dull making a heartless being. Even if leucocytes eat enteric bacteria, they become unable to digest, and it is likely suddenly one day skin eczema comes out throughout the body. Parents taking cold drinks and foods (cold food poisoning) often abuse their children. I want Japanese paediatricians to be aware of the characteristics of the intestine in the infant of the vertebrate just as the American doctors having been aware of for 20 years. Japanese doctors are too idle.

Wrong Obstetrics

During the latest 20 years, caesarean operations have become popular at unprecedented rate in human history. Though this method has been performed since Cesar's time under unavoidable circumstances, it should not be so easily used as often as today.

So as to do spontaneous delivery, you have to activate internal respiration by phototherapy and train the muscles or organs of the pelvis and its surroundings by doing gymnastics. If a child does not pass through the birth canal, he will be weak in the skin and the lungs. As finishing development at the embryonic stage, it is necessary to finish the skin by tightening the body in the narrow birth canal, and at the same time to make amnion fluid stored in the lungs discharge completely out of the mouth. Conversion from water-respiration to air-respiration does not go well until a little amnion fluid remained is absorbed by the lungs. The swell of peristalsis of the intestinal organ such as the vagina and uterus beginning at the peristalsis in reproductive behaviour at the time of fertilization finishes by delivery.

Friction by a baby's birth canal passage creates the unequalled pleasure to finish a sequence of sexual desire, and serenity with the state of Oku or Nirvana restores. Since a mother's delivery has not finished in caesarean section, the mother falls into dysphoria just like sudden interruption of sex intercourse, which gives rise to acceleration of insatiable sexual desire to leave trauma in the mother leading to destruction of home. Since the uterus is part of the intestine, it is naturally an organ in which heart resides. A baby should not be taken out violently by cutting. It should be taken care of that a mother whose uterus has been injured may be obsessed with sexual desire after delivery, and even after menopause to do lewd deeds without thinking of children or home.

Medicine In Organ Transplants

Now that it has been cleared that transplanting internal organs by pulling out as recognizing a living man as brain death is the same as transplanting a heart, as shown in a memoir of Claire Sylvvia I can't help confessing that brain death is wrong in terms of the law of nature. The time has come to develop hybrid-type organ (complex type using transplanted human genes) after ceasing allogeneic transplant, or to put the organ transplant method making use of a shark's (a primordial vertebrate) tissue clarified by me into practical use. The success of transplanting the brain, nerve, cornea, skin, muscle, cartilage, intestine, liver, and so on of a shark into the same organ or tissue of an adult dog or mouse (Fig. IV-1) means that a primordial animal is in the same condition of immune tolerance with an embryo of a vertebrate, and it means the same with embryonic stem cells. Our ancestors have really been Heterodontus as our ancestor's shape shows.

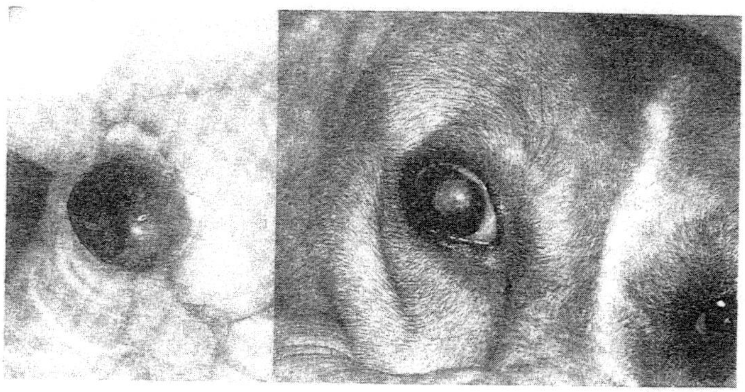

Fig IV-1. Cornea transplantation from a shark to a dog. (Left: Eye of a shark. Right: Successful transplantation to a dog)

The Kojiki in our country teaches us that Toyotamahime, a grand mother of Jimmu Emperor, who was a mother of Nagisatakeugayaaezunomikoto who was the father of our Imperial Ancestors, Wakamikenunomikoto (Kanuyamatoiwarebiko*Inami Jimmu Emperor), was Fuka (a great shark) as big as eight fathers. Though this story was created in three thousands years before Christ, if you multiply it by ten thousands times and fast forward to 3 million years ago, you will notice that since the Devonian period our ancestor was fast Heterodondus, transmission of the Kojiki has been attained by the visceral brain of the limbic system, namely, by the intestine. The Yamato race had known by the visceral sensation of the intestine that our ancestor was a great shark. This is life memory which cells of the intestine have, and culture of hara. Committing hara kiri is considered to be based on sensitivity of our race, or keeping an open mind, based on our culture.

Over-Pursuit of Pleasant Stimuli by the Development of the Cerebrum

A living organism has no object by nature. A reaction in electric phenomena within the finest water-soluble colloid on the earth floating in the universal space is life. Therefore, higher living organisms can continue to survive idly even for 500 million or 600 ones without changing behavioural patterns, given a constant environment. They do not like change. They want to avoid accepting a new way of living as much as possible. When they are in the fittest condition, living organisms are in the state without comfort or discomfort. This is the state of Oku that forgets the existence of the hand or leg, does not notice respiration or hunger and forgets the existence of the body. This state of Oku is the most pleasant state for each cell composing the living organism having no purpose. During this period, remodelling has been carried out. The operation of remodelling provides each cell with unequalled pleasure. Namely, cells composing the living thing like most the lukewarm state called Oku.

When you acquire the movements of your body in sports or driving a car, you are aware of them in the beginning, but soon later your muscles come to remember their rhythm unconsciously, which is called "memory has been established". The real meaning of memory is that muscles learn rhythm, by which you will be able to measure and think. When you encounter a new event and begin to employ muscles, and repeat several times, you can do unconsciously at once. This is memory.

At this time, a new way of thinking, or how to employ muscles has been accomplished. The development of muscles and that of vessels accompanying the sympathetic nerve, jumped on terrestrialisation to generate the power of thinking.

In primordial living organisms such as cartilaginous sharks, they don't have the blood vessel system to nourish the brain and internal organs, but after landing the pyramidal system nerve of the brain, the sympathetic nerve, and blood vessel system have greatly developed. It is human beings that reached the peak of the development. By over-development of the cerebrum, vision-centred image culture has rapidly progressed in the recent decade. And by over-stimuli, the state of Oku, or the lukewarm state that is neither comfort nor discomfort, has been forgotten in our daily life. The state of Oku for each cell of the human body can be obtained during sleep or bathing when gravitation has been released from attraction of the earth. At this time each cell has become in the state of Oku to remodel and enjoy.

Five desires of property, fame, sex, appetite, and sleep becoming the source of heart residing in the intestinal organs are contracted into three desires. The first is appetite, which is the source of possessiveness. The next one is sexual desire, which leads to desire for fame. Sleep is the desire of remodelling combined system of the parietal and visceral, and is the desire similar to muscle sensation at convalescence after sports.

Sexual desire has originally been the desire of remodelling an individual as a whole. Therefore, if one gets caught in this as the ultimate desire, there appear often some who lose everything in exchange for life. When you devote yourself to sexual behaviour, every cell of your body is immersed in joy of the state of Oku. Today's human sexual desire is completely separated from reproduction, and pursuit of joy plays a key role in the act. Sports or religious practices for asceticism are also pursuit of joy. Once or two times of sexual excitement in a year in the general mammals had developed into once sexual excitement in a month in the human beings as estimated by menstruation together with the formation of civilization since several hundred thousand years ago by superfluous nutrition accumulated for 5 million years, and the trigger of reproduction has shifted from the sense of smell to that of sight. In today's visual image culture, pursuit of joy by seeing directly connected with sexual desire has caused a serious problem in civilized society.

Now that it has been clarified that heart, soul, or spirit is substantial life energy and in which visceral intestinal system heart resides, as I have shown in this book, we have to think profoundly "What is life?"

and think of what is sexual desire being the desire of the heart; A new discipline of living has to be created in order to recover smooth balanced life based on five desires, property, fame, sexual desire, appetite, and sleep.

Life is "the system to remodel itself together with the rotation of the energy vortex". Its essence has been maintained as heart equipped in cell itself since the monocellular protozoan, and has been succeeded to human beings of the mammals. Its heart exists mainly in gut organs. Since gut organs have five desires, property, fame, sex, appetite, and sleep in searching for remodelling themselves, we must endeavour to keep always the five's balance smooth. If you admit other's smooth five desires, everything might go well smoothly. This may be the rule for living of human beings who have several structural defects, though reaching the summit of evolution having heart and spirit.

Recognizing five human structural defects, you have to conquer them. You have to know how important energy with no mass is in our life. And correct wrong medicine as well as correct the wrong view of life. If you correct your way of living by accepting energy correctly, you will be able to avoid imminent crisis. In order to do that, you have to change your way of living.

Chapter Five

Face and Heart, Body and Spirit
In the Beginning

I should like to start this chapter with the view point of "Face and heart, body and spirit". In this chapter, as achievements of phylogeny until chapter 4 will be rethought from this viewpoint, I want you to allow several repetitions.

It has been said from the old days that the face represents heart. The source of heart resides in the movement of the intestinal visceral smooth muscle, or in the swell. Namely, it resides in the tremble or twinge of the intestine, lungs, or heart. What the West is to culture of the brain, our country is to the gut. Thus, in our country the gut has been said to represent one's senses. The source of life is placed on the gut.

The essence of living organisms is remodelling (metabolism) occurring together with the rotation of energy vortex. This conquers aging. Remodelling an individual as a whole is heredity, which is carried out through reproduction. Therefore, the most important of living organisms is reproduction. Reproduction is self-reproduction on a scale, and is the ultimate of self-realization in the human culture activities is also spiritual self-reproduction on a scale quite similar to reproductive activities. In the gut resides ego, mind, and heart, and the original organ of reproduction in self-reproduction on a scale lies in the gut, and reproductive activities are performed by the gut.

Assuming that heart lies in the intestinal visceral system, how would the face relate with the heart? The face is called the visceral cranium in medical terms, because the facial skeleton and muscles have been originally moulded by the visceral organs called "the branchial intestine". In the process of evolution, cartilages in visceral skeleton and smooth muscles in the intestine have changed into bone and stratified muscles, but they are clearly derived from internal organs. All the muscles of the face ranging from muscles of expression, mastication, to auditory ossicle have been derived from the visceral muscles of the branchia, and the tongue in the mouth has also been the visceral smooth muscle of the branchial intestine. When you eat the tongue of beef, you feel quite different taste from a round of beef, because the tongue was the same respiratory muscle as the cardiac muscle.

While conventionally the tongue has been said to have derived from the same parietal stratified muscle as a round of beef, the author has clarified that it was the respiratory muscle by dissecting many sharks or amphibians. Originally, these muscles have been innervated by the nerves of the extrapyramidal system exiting from the internal organ being the old cerebrum existing since the age of an ascidian, but in the process of evolution they all have come to be innervated by the nerves of the pyramidal system of the cerebrum. Namely, they transformed into the muscles moved by the power of will like the stratified muscle. The pyramidal system is the voluntary muscle system, and the nerve exits from the cerebral cortex. This is, in fact, the nerve and muscle system that generate spirit and thought. Therefore, the face represents heart and mind simultaneously. This combined expression is often recognized as the personality of the person by people.

The face is the intestinal tract having skeletons, and outside the skeletons appear muscles. This skeleton undergoes changes in shape by external force. This transformation law is called Wolff's law, but, to be simple, bone becomes dented by being pushed, and elongates by being pulled. Gravitational force has been completely forgotten in modern life science. If you are awakened to gravitational force to introduce this law, you will easily understand the enigma of the development of the deformation of the face. Even the enigma of this deformation about the reason why it develops had not been solved until the 20th century. In this chapter, I'll describe first of all the functions of the face, and the method to cure deformation. After that, though a little bit scholarly, I will describe the relation between the face, heart, spirit, and the organs of the whole body.

Function Of The Face

Conventionally, the face has not been thought as an organized complex organ. The face has not been recognized as an important organ, and nobody has ever thought out the function of the face or organ specificity. When you think of "What is the face?", it is the safest to trace back the origin by tracing phylogeny. This is the method of morphology that Goethe created. Namely, it is the surest way to learn from evolutionary science in order to solve organ-specificity. Tracing back the evolution of the face, we reach the head branchial portion of a shark and cyclostome, and further reach an imago of an ascidian consisting of mouth pouch having gill pores.

An ascidian is a mouth pouch with gill-slits, in which branchial portion it has haemopoietic nest, the kidney, adrenal system, pituitary gland, heart, intestinal duct system, and the reproductive system in disorder. It is a sac of life itself. (Fig.V-1)

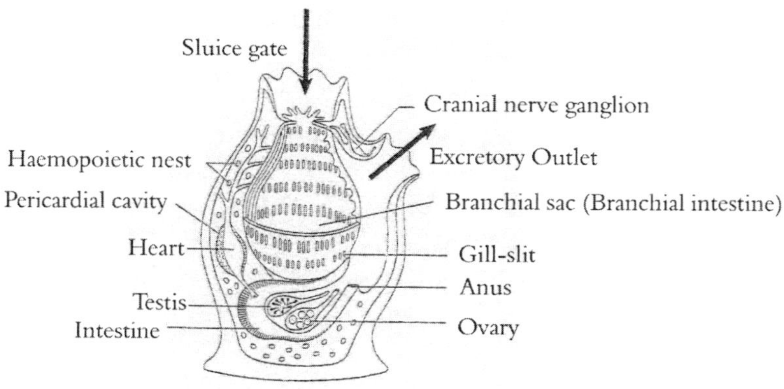

Fig. V-1 Body constitution of an ascidian

There's a proverb that the liver and kidney are the pivot of life. However, the branchia is much better than the liver. An ascidian becomes chain-salpa type tied in a row by gene duplication, and when it does cephalad movement after birth as an individual, the first ascidian becomes the face. The vascular haemopoietic system moved by accompanying the gill movement of an ascidian becomes the heart derived from the branchial gland. The nerve system of an ascidian consists of the brain in charge of smell, sight, equilibrium, and branchia, and the autonomic nerve (the parasympathetic nerve system alone). Predation and reproduction are under the leadership of the sense of smell, because they are dependent on chemical substances solved in water in the water. Namely, it is in the prototype of vertebrates that reproduction has been triggered by the nose. An individual constitution of an chain-salpa type's multisegmental ascidian changes by cephalad movement to Acanthodii, cyclostome and cartilaginous fish, and further by landing to amphibians, reptiles, and mammals. Namely, the face is one part of four parts, head, branchia, chest-abdomen, and caudal portion, into which multi-segmental ascidian in the primitive type had differentiated by cephalad movement, and further in higher animals one part of five parts, face, neck, chest, abdomen, and hips, which is the organ representing life.

Life consists of two phases of food and respiration. The function of these two phases has converged on the face. Even in mammals, generally, the nose is in charge of the trigger of reproduction. Jacobson's organ (a kind of smell organ in the nostril) is the functional effect organ of reproduction. In the primate ascended the tree, the sense of the smell has failed and in humans this phenomenon is prominent above all. In humans, replacement of Jacobson's organ has been performed by the sense of sight in place of the sense of smell. Particularly, the males have changed in this way. The brain develops from the same ectoderm with the skin. Every sense organ, the eye for sight, the nose for smell, the ear for hearing, the tongue for taste, the skin's nerve for touch, is the protuberant organ of the brain.

Viewed from the points of function, the tongue, touch, sight, hearing, and smell are all equivalent. While the trigger of reproduction in dogs is pulled by the smell at Jacobson's organ, It is pulled by the female's looks (especially the face) in human beings, because of the shift to the sense of sight in males. In the Islamic world, because of this, only adult females wear chador. The reason why the females in many countries put on makeup by putting on lipstick or powdering their faces may be because the face is the effect organ for reproduction. I enjoyed the Islamic strip-tease show, when I visited Turkey for the fieldwork of the face. In the show, the belly dance dancers with their faces covered with veil removed the veil at the end and exposed only their face to the spectators. Males are attracted sexually mainly by the sense of sight, while human females are said to be mainly attracted by the sense of smell, touch, and heart (appearance of visceral sensation). Generally, females preserve the primitive type. The trigger of reproduction has been pulled by the parts of the human face having the most complicated and highest function, and at the same time the face has the function to express spiritual life.

The trigger of reproduction has originally lain in the sense of smell of an ancient ascidian, and even at the stage of a shark the sense of smell is protuberant part of the brain. However, human olfactory sense has become so small that a thin nerve from the brain has distributed around the most superior concha of the nasal cavity (the eave separating the human nose into four parts, in which the olfactory nerve distributes) through the thin Lamina cribrosa of the cranium. But, the olfactory nerve is the oldest cranial nerve, which descends into the nasal cavity without crossing with the nerve in the right side descending in the right nasal cavity. Thus, the olfactory sense is the nerve of the oldest and important visceral intestinal system, and lets the intestinal visceral system know the conditions of food and reproductive environments in the outer world. It connects with every visceral organ at the central site, coupling the body movement with the intestinal movements of predation and reproductive system.

The immune system is "digestion, metabolism, anabolism, catabolism, and excretion at the cellular level", and refers to the whole area of life process in charge of the mesodermal system cells. From this point of view, the function of the nose is known to back up the essential function of immunity by controlling respiration and hormone. If you want to be healthy and handsome, you have to activate your internal organs by food and sexual aroma, and create life filled with the fragrance to make the muscle system vivid.

Anatomy Of Expression

In the next place, I describe expression, namely, about mentality and the face. Muscle groups concerning expression, mastication, swallowing and phonation are derived from the branchial muscle. (Fig V-2)

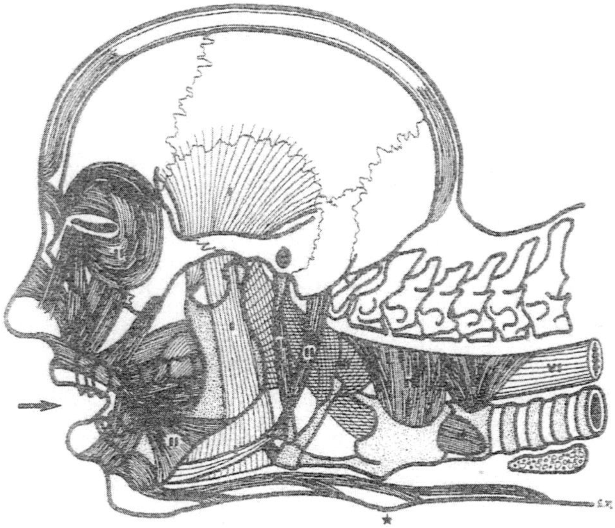

Fig. V-2. The branchial arch muscle of a shark and muscles of transformed human face.(Original bt Shigeo Miki, [The work of the internal organ and the heart of Children,] from Tsukiji Shokan)

Let's think of what the branchial apparatus is. The character of 鰓 (gill) is composed of 魚（fish）, 田（rice paddy）, and 心（heart）. 田 represents the figure of a transverse section of the brain of a man or monkey, and 心 represents the intestinal visceral organ by represented by the heart. The heart is the haematopoietic organ moved by being accompanied by the rhythmic movement of the primordial intestine. The heart is subordinate to the intestinal ducts. Judging from this, the branchial apparatus turns out to be the organ expressing the essential soul of life, or "thinking", in which spirit being the brain function in a fish and heart being the function of the visceral intestinal system are integrated. A fish represents one's feelings by the gill.

The ancients that created Chinese characters, had observed living things very accurately by the limbic system of the visceral brain. Originally, heart and spirit have been separately handled regardless of the West or East. Lately, medicine and biology have confused, and these parts around have been jumbled up. However, heart derives from visceral intestinal sensation, while spiritual activities derive from the parietal muscle sensation. From the first, visceral sensation passes through the ciliary body of the brain stem to enter the visceral brain, becoming the linchpin of emotions. Spiritual activities are carried out with these emotions being the core. Soul is what the visceral brain and the spiritual brain the consciousness of the parietal system united, namely, is inseparable life itself in which the body, nerve, and heart have united. What has become of the branchial intestine and branchial gland (smooth muscle and hemopoietic nest) have become the muscles of the face and tongue, and diaphragm in human beings, and at the same time become the Waldeyer's lymphatic ring, thymus (both: leucocyte hemopoietic organ), lungs, and cervical sinus (originally erythrocyte hemopoietic organ at the branchia). The mammals that employ respiratory muscles of the branchia except the diaphragm as other movements, use the parietal pectoralis muscle, lattisimus muscle of the back, straight muscle of the abdomen, and lavatory muscle of the anus to move the lungs and respire, by making hollow or inflating the breast and abdominal cavity. But today, movements of the branchial arch muscles such as expression muscles in connection with respiration can be seen only in a yawn, a sneeze, and nasal alar respiration (substitution for lung respiration by moving nasal alars) at one's death.

But, the old branchial arch muscles are linked well with today's parietal respiratory muscle groups. Since the expressive organ of one's feelings is the branchial intestine being the respiratory intestine, laugh being respiration of the intestine can be done by using the muscles of the face, abdomen, breast, and anus (as a dog wagging its tail). Then the pituitary gland or adrenal gland associated with the branchial apparatus shows positive reaction with pleasure. Accordingly, laugh has the power to cure illness. Grief is of negative respiration, with which your heart burst, or you split your sides, while naturally your adrenal gland will have bad blood circulation, which turns out to bring about diseases.

Difference between right and left of the face, and deformation

The character of "色" meaning complexion has also the meaning of sexual desire. The ancients created characters by the limbic system; they created "色" by the form of sexual intercourse, and at the same time applied the meaning of "顔" to the character. The functional and effective organ of reproduction has transformed from the nose with pheromone to "色" of visual sensation, and for that reason sexual desire that was colourless and clear has had the tone of a colour.

Human beings are different from the other Eutheria (mammals having the placenta) in that the formers could have afforded to store superfluous nutrients and power in addition to having acquired nutrients and safe sleep by various contrivances. By using this superfluous power, man expresses dynamic life through the manners suited for an individual. This is culture activity proper to humans. In the days of childhood they are called study, sports, or plays, and in the adulthood they are called task or leisure. Through these activities man attains self-realization willy-nilly. The first step for daily self-realization is making up common to man and woman, or toilet.

The face can be asymmetric, which is due to difference in function. Many of the right-handed person are right-masticated. The left half of the face is absent-minded (spell-bound) with the parasympathetic nerve predominant, and the right half is of active type with the sympathetic nerve predominant. In the many photographs or pictures of the female the left side is drawn due to the spell-bound face. Regardless of the East and West, human beings are more right-handed. But the faces of saints are symmetrical. No difference between right and left is worthier. Leonardo da Vinci, seemed to have removed the difference between right and left by writing the character reflected in the mirror by his right hand. Difference between right and left interferes with art activities. In today when the secret and enigma of the face have been solved, you can improve your health by improving life habits and removing the difference between right and left by training.

Self- realization in a day begins at the morning make-up, and finishes by washing one's face in a comfortable bath or dissolving gravity. One day finishes by bone-rest with satisfied sleep. Overwork in which superfluous power disappears does not lead to life full of energy. Symmetrical one's face and figure are sustained by right sleep, and comfortable meal accompanied with right discipline. At the same time, health is maintained, restored, and reinforced, which leads to fulfilled self-realization everyday. It is in the face that the brightness appears.

The Reason for the Asymmetric Face

In the human beings, the distinct asymmetric jaw and face has appeared since Homo Sapiens whose dermal bone became thin. The reason is that the change of form by the influence of biomechanics caused by difference between right and left by body movements appears only in the case that bone is so thin to be subject to external force. In the jaw of Neanderthal man, the habits of one-tail bite and sleeping with one side being have been apparently memorized. The dominant jaw and hand are evidently drawn in the right side of the face and upper arm in Leonardesque balanced anatomy of the human body.

Bone has the biomechanic property to change to the form fitted to the function by the inclination of the function. This is Wolff's law (an empirical rule) proposed by clinical research in 1882. The frequent functioning of bone causes bone tissues' diminution, and increases its density. But bone elongates when pulled. Since bone has no system to move by itself, the influences by muscle power and external force given to bone are almost equivalent.

The external force becoming the cause of deformation of the facial skeleton is lumped together under "the habits of the oral cavity and its periphery". These habits are ①oral respiration, ② unilateral bite, ③ sleep lying on one side, or lying on one's stomach, ④ rest one's chin in one's hand, ⑤ clenching one's teeth, ⑥ slinging a bag over one's shoulder, ⑦ playing any musical instrument, and so on. Among them, the face is for the most part deformed by the following three habits such as oral respiration, unilateral bite, and bad sleeping posture. Since the degree of deformation is not even, naturally the difference between right and left is recognized.

Since these habits are linked one another, caused deformities are almost typical. But, if the habits are amended, the deformities will be corrected. The reason for linkage is that the muscles in the oral region are derived from the visceral smooth muscles of the branchial intestine, and the remnants of peristalsis, so that these muscle groups link one another.

The only characteristic organ to define the vertebrate is a tooth of gomphosis with the joint (the periodontal membrane) having the cementum. The tooth is peculiar of the mammals and effective for mastication. But it has the mechanical property that it can not bear lateral force and staying power. A tooth sinks or moves laterally by staying power and lateral force having one over five hundred to one over two thousand of masticatory force. Then, only the tooth does not move, but also the shape of the jaw changes gradually. The tooth moves by staying power and lateral force caused by habits. Oral respiration causes not only buck teeth (prognathism), but also progenia (opposite occlusion), or open bite (nonocclusion).

The mammals have the system to swallow under negative pressure in the mouth. In oral respiration, you obstruct dentition by the tongue instead of closing the lip. At this time, pressure to push by the tongue ranges from 40 gr to 70 gr, and the way of collapsing is determined by the site at which the tongue touches.

On the other hand, oral respiration links only the habit of unilateral bite. In unilateral bite, the chewing side shrinks and is tightened, while the non-use side becomes loose. In terms of chicken, a domestic-chicken-face is to tense what a broiler is to loose. Both the cervical and thoracic vertebra develop scoliosis by bias of the muscles. Unilateral bite links the habit of sleeping posture. Lying on one's side or on one's stomach with the chewing side down is general, by which sleeping posture dentition intensely warps, and the backbone also curves. Under these sleeping postures, the underlying nasal cavity is obliged to be obstructed by congestion of veins, and naturally oral respiration develops. Namely, linkage returns to the first state. If you continue to breathe through the mouth, to bite unilaterally, and to sleep slovenly for 30 to 40 years, your body and mind will not retain the shape, and become a worn-out man or woman bent almost double with age. It causes without fail serious immune diseases. The bone marrow has the haemopoietic organ of erythrocyte and leucocyte in charge of the immune system; Latent infection of the distorted bone marrow by bacteria that enter by the habit of oral respiration causes immune diseases. Difference between right and left, namely, deformation has the influence on the face to distortion of the skeleton of the whole body.

The Way to Cure Difference Between Right and Left

Since the principle to cure difference of the body between right and left is the same as that of the face between right and left, curing the face is to lead gradually to correcting distortion of the body. The reason is that the body distorts by linking with the habits of the face. The way itself to cure distortion of the face is unexpectedly simple, but when it comes to cure by correctively training the employment of the facial muscles, it is extremely difficult, because the habits derived from the visceral muscles have to be reformed.

Correcting effectively sleeping postures, unilateral bite, and oral respiration at the same time, necessitates determination. It is the determination to correct the wrong employment of the body in order to live a healthy life through your life from now on. At the same time you have to cure bad habits obliged to take on using biomechanical devices called a respiratory set of 5 pieces in order to correct three bad habits impairing your health. The followings are very effective for this: a nose-lift made an 18-carat gold to make the nose high-bridged for helping breathing through the nose which the author developed, a fluffy pillow, and a breath-trainer for teething for adults. The others are a lip-tape for becoming pretty during sleep, and an instrument to make the face slim; By the set of these three 5 pieces, the swelling of the face subsides by one week, dark skin becomes white, rough skin or atopic dermatitis improves, and asthma, bronchitis, or interstitial pneumonia improves unbelievably.

The simple treatment for deformation is in the first place to perform gum-therapy with oral respiration and unilateral bite being cured and sleeping posture being corrected at the same time. Since in oral respiration the upper lip and lower lip are not so long as they are expected, all you have to do is to pull them to elongate well. You train orbicular muscle of mouth by pulling up and down, and right and left by inserting your fingers into an angle of the mouth. As for gums, you do not chew strongly, but chew softly by closing or opening the jaw widely, with straightened your spine, and with lips, urethra, and anus tightly closed. You have to train to bite at the reverse dentition against unilateral bite, and at the same time stretch out your neck by bending it toward non-biting side. In daily life or during meals, you have to straighten up, lower your lingual bone with your chin drawn, do diaphragmatic respiration with your stomach lifted, and sometimes stretch your self so as to straighten up. Usually you train well the opposite hand to your dominant hand. In the posture of doing word-processor, since the shoulder in the dominant hand side lowers, you have to correct your posture by bending your body in order to remove the difference between right and left.

The difference between right and left in the face and in the spine is gotten rid of by them. You sleep with fluffy pillow with your hand facing right above. You, adults, have to sleep at least more than 8 hours and have bone-rest. If you neglect to have enough bone-rest with oral respiration, you will suffer from immune diseases.

Relation among the Face, Heart, Spine and the Whole Body

Development of the pyramidal system dominance of the facial branchial arch muscles

Our face is a complex organ representing life, of which origin is an ascidian of protochordate consisting of a mouth-pouch having gills. The mammal is the multi-segmental animal of which the branchial intestine of the oral cavity and pharyngeal region extends to the region of the pelvis: the external skeleton of the mouth is covered by the masticatory muscles derived from the branchial intestine of the visceral muscle, and the face is also covered by the facial mimetic muscle derived from the branchial intestinal muscles.

The tongue that exists in the centre of the mouth is composed of an aggregate of the branchial arch muscles and the branchial arch cartilages in the primitive cartilaginous fish and

At its base exists the heart. The almost immovable tongue of a shark comes to do extremely subtle movement by the government of the pyramidal system (the cerebral mouth nerve) and the sympathetic nerve at the time of the secondary revolutionary landing drama in higher animals. Especially, the human tongue becomes the central organ organizing the sound of languages that convey the state of mind most difficult to express ideas, thought, and spirit.

On territorialisation of the second revolution, the sympathetic nerve has developed with the new birth of capillaries, and at the same time the pyramidal system governing the diaphragmatic muscles has come into existence from the cerebrum. At the birth of mammals in the third revolution, spiritual and mental activities have developed so that in the end man has come to speak. The most characteristic point as a man is language, and this is performed by contriving the employment of the tongue and the oral cavity, which leads to the tremendous progress of the brain. While the tongue in the primitive type moves barely, it comes to move by the hypoglossal nerve of the 12th cranial nerve together with the sympathetic nerve developing at the same time of retraction of the branchial arch cartilage. Since in the face the branchial arch muscles being the visceral smooth muscles have become the stratified muscles, these ones are naturally subject to the combined influences between the extrapyramidal system (the primitive motor nerve in the brain) and the pyramidal system. Half parts of the muscles of the branchial intestines composing of autonomous involuntary muscles originally become voluntary stratified muscles, while the other half parts come to maintain automatic motion like when masticating or swallowing. Thus, the facial muscles are doubly-innervated: the extrapyramidal system of the primordial cranial nerves, and the pyramidal system of the cerebral new cortex of mammals.

Relationship Between the Face and the Various Organs of the Body.

To say nothing of "The principle of the correlation between organs" and "The principle of dependence" by Cuvier's comparative anatomy, all organs constituting the individual of the vertebrate and their functions are intimately related. In order to know the relationship between the face, the comprehensive organ representing the human life, and the other various organs, you may well study the following seven points:

1. the face and the reproductive system.
2. the face, the limbs and science of reproduction.
3. the reproductive system and diseases.
4. the face and the brain.
5. the face and the cardiopulmonary system of the heart.
6. the face and the abdominal intestinal visceral system.
7. the face and spirit in the parietal muscle system.

The Face And The Reproductive System

Since the essence of life depends upon remodelling occurring together with rotation of energy's vortex, the function of the intestinal ducts to maintain it, that is, respiration and eating are most important for living organisms. Therefore, the nose and mouth that are the entrance to respiration and food, namely, the face being the visceral cranium turns out to be the most essential organ. In reality, if your breath stops, your life will end, and if you can't eat anything, you will die. Foods eaten are digested and absorbed in the intestines to become blood and flesh by and by, and superfluous nutrients become fat cells and reproductive cells that are a kind of blood cells at an appropriate time, which are stored in the body.

Since living organisms are maintained by conquering aging by remodelling an individual, reproduction is just the most important for life. The oral cavity being the source of the face that is the entrance to the intestine, is directly connected to reproduction through the intestine that is most important for life. Living organisms live originally without purpose. However, the structure and function of a cell are made so that it feels unequalled pleasure in remodelling (metabolism or reproduction) from the times of a protozoan, and they are memorized to be transmitted to the next generation, the generation after the next, and successive generations. Procaryotes (bacteria) also have sex, and prokaryotes of a single cell are willing to mate.

Genetic materials memorize not only the ways to create the substances such as proteins, polysaccharides, and phospholipids that make the structures of living organisms, but also they memorize chemotaxis (the movement of an organism toward a substance to take it into) for the specific nutrients, oxygen, or chlorides involved with the functions.

This chemotaxis is a kind of energy form, that is, the function of like or dislike. Since like or dislike is the start of mind, a mono-cellular animal already has mind. Therefore, a leukocyte in our body that holds the figure of a protozoan three billion years ago has naturally mind. The mammals that reached the climax of the evolution of the vertebrates, have cells that have differentiated into uncountable numbers. In extremely deviated differentiation such as the bone, tooth, nerve cell or muscle cell, the function that is the source of mind (like or dislike), has been occasionally missing. Here, you have to take into consideration, the intestinal epithlium, sperm, ovum, leukocyte, or histocyte (mesenchymal cell made of leukocyte). These are the oldest type cells. The sperm and ovum are oldest, in which nucleic acid undergoes meiosis and is haploid (mono-helix); they are near to protozoans, and the nucleus of sperm does not have even mitochondria. Mitochondria keep such old shapes that they exist only in the tail portion of the sperm to be cut down away. The sexual desire (the heart for liking the ovum) that the sperm cell has is strong; at sexual intercourse several hundred million's sperms begin to run straight for the ovum; and the heart for seeking the sperm by the ovum is also intensely strong.

In the human being, 500 ova are prepared through its life. Oestrus has changed from twice in a year , the rule of the mammals, to once in a month or twelve times in a year; this is presumably due to the results that nutrients have been obtained stably since several million years ago. Every month mucous membrane of the uterus thickens for fertilization, and ova are formed from superfluous nutrients; just like the same with the case of a hagfish, the ova mature in the abdominal cavity, and come out into the uterus through the oviferous duct by swimming. Then, the ova wait impatiently for the sperm.

When the seminal vesicle is filled with the sperm in the male, the sac of the excretory intestine called the seminal vesicle becomes impatient to peristole. When the cell-groups called the sperm come together in the seminal vesicle, they rouse a whole of individual consisted of sixty million cells. It is the brain that becomes the window between the excretory intestine and the outer world of an individual, and the excretory intestine at the anal side plays an active role. When the excretory brain operates, suddenly the belly feels fuzzy; spirit and mind by the branchial brain, and ethics in daily lives, paralyze to change one's character. Many priests or monks who preach the way, scholars, philosophers, or literary men, would go to hell by straying from the path of righteousness due to the desire of the cell called the sperm. The ovum comes out into the uterus by swimming the abdominal cavity every 28 days, and this is really the remnants of ebb and flow of the tide. And still, this is the remnant of the reproductive formula of the coelenterate living in only the intestines being the source of multi-cellular animals. The coelenterates discharge ova and sperms all at once twice in a year at high tide. The remnants in 7~8 hundred million years before Cambrian period have been succeeded up to the mammals.

The ovum, differing from the sperm, comes out regularly every 28 day in the humankind in the contented condition of nutrition, and comes to want the sperm extremely at a certain age. When manners and customs of sexual behaviour corrupt and pornography is overlooked like in our country or U.S.A today, the students of the elementary , junior high, or senior high school come to experience a sexual awakening. Nowadays, it is in Japan and U.S.A that people, regardless of age or sex, are crazy about reproductive behaviours.

The living fossil developed at the Cambrian period, Lampetra (lamprey), clings to the rock, and the couples perform reproduction forming a line side by side. Shortly after reproduction, sudden aging at the cells of the whole body occurs, and by apotosis (death-program incorporated into the genes of cells) the trigger of the genes to death is pulled. The human beings having undergone the evolution of the vertebrate in response to mechanics for about 500 million years after the birth of an ancient lamprey, suck in intensely the lips in today's reproductive behaviour. The facial muscles are visceral in which the intestinal muscles protruded outside the cranium constituting the anus or face derived from the dermal bone. The testicles and pudendal lips are also covered with smooth muscles in which muscles of the portion of reproductive visceral intestine of the branchial intestine protruded to the subcutaneous tissue of the ectoderm. That's why this portion moves slowly without reason. And at coitus the entrance and exit that become the source of remodelling of the whole individual, namely, the oral lips and lips of pudendum seek strongly profoundly each other not to separate. When reproduction turns sour, one may choose to carry out a suicide pact though he does not need to die, because the basic program of life of an ancient lamprey 500 million years ago may operate. Since the essence of life is the behaviour of remodelling or reproductive acts, one may choose death, if the bonds of society get entangled after finishing the essence of life.

In the primate, the trigger of reproduction has shifted from the vemeronasal organ (Jacobson's organ=smell receptor of pheromone) to the visual sense. Today, in the human being, pornography has lost its control in the world. Disorder of sexual behaviour ranging from Kennedy to Clinton in U.S.A means conceivably the complete loss of sexual precepts in the Christian society. It is high time that control of sexual desire separating from the human reproductive behaviour under the new precepts based on life evolution were established.

At the temple of Hadrian in Ephesus, the ruins of ancient Rome in Turkey where I visited to do fieldwork of the study of the face, there was a large bathhouse opposite to the soaring library, where large chair-type toilets made from marble were lined, and beside it, a prostitute house like a large red-light district had been built; a library and a prostitute house were in a set. Reproductive activities and spiritual ones in the library have been performed by the set in the age of Rome. The reproductive act is the remodelling of the whole of an individual, namely, regeneration and reproduction on an expanded scale of an individual. The cultural activities that only human beings have, are spiritual-productive, are self-realization, being the nearest desire to reproductive activities in terms of producing things. Writers or painters, regardless of male or female, tell that the joy they feel at the completion of good works is the same as the satisfaction when they feel at the best reproductive act, with every cell of the body being joyful. Thought activities in the library is the ultimate self-realization, that is, spiritual reproduction of self on an enlarged scale. The visceral brain and parietal brain are there united to act as the extremely similar living individual.

I've stated; the urinary-reproductive organ is the mechanism to intermediate between the external respiration of the branchia and lungs, the intestinal work of nutritional absorption, cell respiration and energy metabolism of mitochondria. Since we prepare well oxygen and nutriment absorbed in the intestine being the source of life to remodel cells, converting them to fats, destroying old cells, making them urine, changing superfluous nutritious fats into the sperm and ovum, urine and reproductive cells are equivalent for life; distinction between dirty and clean does not exist at all. At the age of an ascidian, the mouth and anus existed at the same upper edge of the sac, and the sperm and ovum went out from the cloaca, that is, the anus alone. This has been handed down from the primitive even to the birds or reptiles. Sexual desire is fundamentally subject to contact of reproductive organs and stimulation of peristalsis.

A human baby finishes also as a mammal through the mammalian-type reptile until two and half years after birth when he finishes as a human baby. It is not until he becomes 24 years old that man's growth and development finish in terms of the real anatomy. Until then, man develops after a lapse of years through the various stages of phylogeny. From the age of an infant to around 10 years old in one's boyhood or girlhood, sexual desire converges on the anus. At the time of the cloaca, the anus contained both the reproductive organ and the birth canal, so that there exists naturally a sexually sensitive area even at the anus of human kind that traced the main road of phylogeny.

Let's consider talk about indecent matters and perversion from the viewpoint of medicine. Rumour had it that a politician called a perverted man made his lover drink his urine or drunk her urine; the urine, or the waste produced from 60 billion cells via blood, and the sperm and ovum being superfluous nutrients are equivalent for the living body, having no concept of clean or dirty. If he or she does not suffer from illness, both the sperm and urine are aseptic and germfree. I side with the politician in one aspect; even if a doctor performs 'urine-remedy' to make his patient drink his own urine by diluting with much water, he does not violate the Medical Act. Let's consider a shark of the primordial vertebrate and a whale or a lion sea of the aquatic animal (the mammal). In the water, they are already performing faeces and urine remedies by thinned their own faeces and urine. By any chance, as the doctor of urine-remedies tells, trace components of urine may really be effective by feeding back. However, this is only applied to his or her urine; in the case where the lover drinks the politician's urine, it is unknown whether it is effective or not.

I've previously stated that the face and reproductive organ have become the trigger of reproductive behaviours by the sense of vision; both the face and external genitalia are originally the system of the intake of nutrients by the branchial intestine and the one of excretion of metabolites, and their origin is the same one. They are governed by the branchial brain and excretory brain; the entrance and exit of the intestine is in fact the reproductive organ in terms of the real meaning of the intestine (real work). The external genitalia and anus have only sex secretory gland as the sensory organ and the sensory receptor of the skin. Therefore, they feel only the sense of touch by peristalsis, while the branchial intestine with the mouth centred contains the eye, nose, tongue, ear and skin

The age of Greece and Rome before the appearance of Christian society where all the people became the slaves of the pope, was so magnanimous that there were no movement against the flow of life evolution. Since 'spiritual awakening' of religion indicates originally having acquired 'What is life?' to the utmost, the teaching of Jesus who is primarily a son of a man, must have been generous, not against the human life evolution.

However, the understanding of the universe 2000 years ago is too far distant from that of today, and many have been resolved. The understanding of life also must be revised in accord with the new constitutive law of the universe. In other words, it is the bright and throbbing pulse of 21st century where the current of Greece and Rome before Christianity and the thought pf a comma-shaped bead (the spirit which is present in words throbs) have been fused. The recovery of the reproductive rhythm based on the right phylogeny in which life is in accordance with the rhythm of the season or age, is desirable.

The Limbs and the Reproductive System

The entrance to the branchial intestine is the mouth, and the face is formed around the mouth. The upper limb is an auxiliary organ of the intake of the branchia, and the lower one is the equipment to excrete the wastes of nutrients taken into by the branchia, its origin being an auxiliary organ; the cartilages of the fin are the anlage of the arms and legs. The legs function at the time of delivery or coitus as an auxiliary equipment of the reproductive organ. There are people men and women alike who become sexually aroused just by looking at the legs. A ball game in which only the legs are used is soccer. 'The World Cup' excited men and women of all the world. Man's movement to carry the attack to kick a ball into the opponent's goal may by any chance similar to a sperm that runs searching for the vulva and an ovum. Moreover, soccer is the only ball game that is played by the legs being an auxiliary apparatus for reproduction.

I'll show few examples of the ground that the legs are an auxiliary apparatus for reproduction. Just as a female has menarche, in a man semen emits naturally during sleep with a dream of a female around the first grade of junior high school. This is nocturnal emission. At this time, he awakes from the dream, and can't walk instantly with the legs paralyzed. An unknown innocent man thinks of having suffered from a serious disease, and rushes into a toilet after a while. He falls asleep after returning to his bed wondering why it happened. Presently, when semen accumulates in the prostate gland being part of the intestine, this stimulates the excretory brain in a dream to evoke life memory, leading to ejaculation with image by the visual sense. In the dream of an honest junior high school student who had not think of a female, the morphology of the female sex organs has appeared; just as Clair Sylvia comes to know in the dream the figure , name, and favour of food or a female of the male donor, the cells of the intestine have remembered all.

The Sense of Smell and Heart

The first cranial nerve, or olfaction closely resembles taste sensation in the water, and therefore it is needless to say that it is part of the mouth being the eating system. The olfactory sense is the oldest cranial nerve in terms of phylogeny, and the only cranial nerve that does not cross after exiting from the right and left of the brain among twelve pairs of the cranial nerve. At the infant stage of the revolution of evolution, or at the stage of an ascidian where branchial respiration (enteropneusta) begins together with predation by the intestine, the end of part of the brain remains to stay at the entrance of taking food. The sense of smell has connections in terms of the nerve with all organs of the visceral intestinal system and the parietal system, and controls hormone-secretion and movements of muscles. In the origin of the vertebrate the very beginning of impression is almost the sense of smell. The origin of impression is to move the parietal muscles so that the intestines may move the mouth and branchia in search for food and reproduction. The equipments for reproduction, absorption of nutrients, haemopoiesis by superfluous nutrients and reproduction, exist at the mouth becoming the origin of the face and the intestine of the branchia, and it is the sense of smell that pilots the living individual according to the requests of the intestine. The muscles moved by the sense of smell, are all derived from the old cerebrospinal nerves, so that they belong to the parasympathetic nerve system and the pyramidal system.

Smelling the azalea waiting for May
I smell the scent of the sleeves of the people in the old days

(Anonymous)

The pasania
The scent exciting also a man (Buson, Yosa)
Resemble the heart of a traveller

The pasania

(Basho Matsuo)

The scent of a flower is directly connected with the sense of the reproductive system: the sense of smell is directly connected with the extrapyramidal system of the muscle-nerve system that controls the fundamentals of the life of the vertebrate such as respiration, food-taking, and reproduction. The reason why Saigon and Basho moved around travelling through their lives might have been the heart to search for the reproductive system and food-taking which the intestine did not stop wanting. Aroma-therapy is effective: it increases the digestive power of leucocytes through hormon-secretion by stimulating the parasympathetic nerve.

Rhythm of Respiration and that of Waves

The life of the vertebrate starts from the intestine in both ontogeny and phylogeny. Since the intestine is the origin of life, life does not die as long as the intestine lives.

Paraneuron and neuron segregate at the outer membrane and the intestine, which develop around the mouth and anus to become the branchial brain at the mouth side and the excretory brain at the anus side. One of the centres of the two parasympathetic nerves becomes the brain of the intestine in charge of respiration, digestion and haemopoiesis, and the other becomes the brain at the anal side to control superfluous nutrients=reproductive cells, urination of waste osmotic fluid, and storage of residue after digestion of food. The brain between them becomes the intestinal brain.

Since the heart and lungs derive from the branchial gland of the branchial intestine, they are the intestines of one's heart, where one's heart dwells. The heart representing the excitement or impatience of the visceral organs expresses joys with the following; "My heart is pounding", "My heart flutters", "My spirits soar", or "Imagine across one's mind",; and expresses sorrow or pain with the following, "Feel as if one's heart would break", "It makes my heart ache", "It wrenches my heart", "One's heart pains one", or "One feels uneasy around one's heart". The impatience of the intestines and heart has evidently the different objects of emotions. One's heart dwells in the visceral intestinal system represented by the heart. One's self dwells in the pitch-dark intestine. Today, many scientists think that one's heart exists in the brain, however, has been thought to exist in the gut in all countries of the world.

The vortical rotation of life energy is a wave motion quite similar to rotatory motion called peristalsis of the branchial intestine. What force has caused the vortical rotation of life energy of the higher multi-cellular organism being the first vertebrate? This is in fact wave motion in the sea. This has become the source of respiratory motion of a hydra (Coelenterata), a lophophore, or Bryozoa (Pterobranchia). If you shake a fainted person out of sleep, he awakes suddenly, because the source of respiration has been fluctuation movement of wave. The repetitive movement of the waves coming in and going out has been memorized by the muscles such as the heart muscles that are the intermediate muscles between the parietal and intestinal visceral muscles, and this has been memorized by the movement of the heart. That's why the rhythm of respiration is coincident with that of the wave (Shigeo Miki).

I mentioned previously, "The vortical rotation of life energy has no purpose", but is it really right? Remodelling creates the greatest pleasure; since the engine is produced by nucleic acid, protein, saccharide, phospholipid, and so on, the cell themselves, of the protozoan composed of one cell or even the multi-cellular animal, should want to become energetic by remodelling. Because of this reason, many animals including a man go naturally to sleep to rest. In an adult as well as a baby, the fluctuation movement of waves such as a cradle is very important at the time of remodelling, growing, or developing during rest, because the root of respiration lies in the movements of waves. It is in the original behavioural pattern that respiration, intake of nutrients, digestion, and absorption are performed by the movements of waves, and reproduction by the ebb and flow of the tide (the gravitational pull of the moon). That is to say, the peristalsis movements of reproductive behaviour in worms or mammals are originally the movements of waves.

There is a novel called "Shiosai" (the murmuring of a rising tide) by Yukio Mishima. He was so sensitive that he felt sexual desire by only hearing the sound of the waves; he could not have imagined that this had derived from the profound fundamental system of life of the vertebrates of 500 million years. Since man is in rut all the year round, he feels uneasy just on hearing the sound of waves. In the animals, sexual excitement reproduction being remodelling an individual as a whole has been controlled by hormones with the gravitational pull of the moon and seasons, and programmed to operate twice in a year by life-machine.

The whole cells forming a living individual feel the greatest pleasure twice in a year. The trigger of reproduction in many animals is pulled by the chemical substance (pheromone) related with the smell. The secretion of this pheromone has been controlled by the seasons or the gravitational pull of the moon.

The mouth, jaw, and intestinal visceral system
Basyo composed the following haiku,
Interesting
But sorrow in a short time
A cormorant boat

This is a haiku feeling ashamed of human cruelty by coming to oneself that man watches the exercise of ordinary life activity by collaring and damming up a cormorant's neck with interest.

The pitch-dark intestine starts after passing through the throat. Inside the abdomen, neither thermal stimuli nor powerful drugs can be sensed. It is in the pitch-dark intestine or in the intestine of abdomen that self resides in. No matter how they love, they can not satisfy their hearts with love, if they are extremely hungry. The vertebrates can satisfy the desires of their selves, only when water-soluble nutrients with mass are sufficiently digested, and reproductive substances being superfluous nutrients are sufficiently filled.

Since the brain starts only from the intestine, the former subordinates to the latter. Accordingly, the brain functions so that it may realize only the wants of the intestine. Neither struggle for fortune nor sexual desire nor desire for fame cannot be controlled by reason, namely, by calculation by the parietal muscle system of the brain and spirit or thought: Five desires dwelling in the intestinal visceral organs are desires that have been taking over life memory of self of 500 million years in phylogeny.

The face and mouth are composed of the branchial respiratory visceral muscles that are what has become of a once branchial intestine in which the sympathetic nerve and the pyramidal system are stratified in terms of innervation. Expression represents heart in the function of the body wall. Since eyes are part of masticatory organs, they represent well both heart and spirit. Language also represents heart and spirit to the rhythm.

Since the mouth has been programmed to use for eating-mastication, too much talking with oral-breathing damages Waldeyer's tonsillar ring (a kind of the intestinal tonsil and leucocyte-hemopoietic organ) derived from the second branchial gland. It is in man alone among mammals that this tonsillar lymphatic ring has developed extremely.

This ring in a dog, cat, rat, horse, or cow is as small as it bearing the traces. In the primitive sharks and bony fish, the second branchial gland corresponds to this, having become an extremely developed hemopoietic organ. It is because they both breathe through the mouth that the ring has developed only in man and fish.

What supply the substances to sustain remodelling and energy, are nutrients by food and oxygen by respiration. Foods masticated by the mouth are converted to blood via the intestine and reproductive cells. Hemopoiesis by the intestine necessitates sufficient mastication. Therefore, sufficient mastication and right nasal respiration turn out to be the source of healthy reproductive activities. Namely, the essence of life lies in the function of the mouth with mastication of food centred, the respiratory function by the nasal cavity, or the functions of the visceral cranium.

The genita organ is a kind of haematopoietic organ and there exist a haematopoietic organ and kidney-genital organ in a lancelet's intestine. The kidney is the system of excretion of wastes produced by muscles and a kind of the branchial haematopoietic apparatus of the mesenchymal epithelial system (mesodermal epithelium). That is to say, the kidney-genital organ is part of the hemopoietic apparatus of the respiratory system, which moved to the anal side with the law of inertia by gravitational force by cephalad movement. It is because of this that breathing through the mouth, not through the nose, causes often infection of the urinary-reproductive system. In females, cystitis, menstrual irregularity, menstrual pain, endometriosis, or sterility may develop, while in male, prostatitis, cystitis, or oligozoospermia may develop. Correction of occlusal disharmony and restoration of nasal respiration lead to cure of these diseases in many instances. Heart has its source in the intestinal visceral system; it is the visceral brain that conjugates with these visceral muscles, which is composed of the limbic system, hippocampus, and thalamus-hypothalamus, where neurons exist to catch the conditions of the intestine. Because of this, yo9u become very eager to see people and have an oppressive feeling when the intestine twinges.

The Face and Spirit in the Parietal Muscle System

After landing in the second revolution, when gravitational function becomes six-fold, life media changing from water to air weighing one thousandth of water, coming to be able to live against gravitational force. Muscles move actively by leaps and bounds, and at the same time cellular respiration of all tissues and organs of the body is activated. The muscles moving especially hard, demand plenty of oxygen; as a result, cytokine to induce blood vessels is conceivably secreted from the cells. As a result, induction of blood vessels occurs. The mesodermal sympathetic nerve system develops in new born vessels as the vasomotor nerve. Nutrient canals develops in all organs such as the heart, brain, and the digestive ducts of archetype vertebrates that have not have feeding canals so far, and at the same time the brain , heart, or intestinal visceral system are doubly innervated by the sympathetic and para-sympathetic nerve. Muscles are functional organs of the nerve. Whether it is recognized (the parietal system) or not recognized (the visceral system), when the afferent nerve conveys information to the centre, electric potential develops in the central nucleus in accordance with the situation, and the information is conveyed to the efferent nerve to move muscles.

The sympathetic nerve develops when the parietal muscle motor system has developed greatly after completion of bipolarization to the mouth and anus of the parasympathetic nerve, and the brain, internal organs, and heart respond to the outer stimuli. Cowering under terror, feeling stresses, and being leaped with joy, or having a strange presentiment by outer accidents come to be felt by the heart and lungs. When the cerebral motor nerve of the pyramidal system developed after development of the sympathetic nerve, respiration and energy metabolism of the glycolysis system (bioreactive process without oxygen to break down glycogen into lactic acid), made rapid progress; change from poikilotherm to hemotherm has occurred, and homeothem has been born. At the same time, the tongue, hands (fins), or legs (fins) came to be moved by the power of will.

Development of the Sympathetic Nerve System and the Pyramidal System, and Homeothermism

The sources of thought and spirit have surprisingly existed in the rhythmic movement of the pyramidal system of the parietal muscle system represented by the dorsal muscles (one's back). The sound existence of the extrapyramidal system governing the old muscles (intermediate between stratified and smooth muscle) of the parietal system unified with the visceral muscles attached to the intestines existing since the age of an ascidian being the source of an archetypal vertebrate, is essential for spiritual activities. The parietal muscles moving together with the internal organs are the muscles move by respiration, eating, reproduction, sleeping and rest-posture.

It is necessary for brainwork also to introduce rhythmic movement of muscles tuned to respiration. This necessitates the healthy pyramidal system being the parietal system of paleocortex of the limbic system.

Generation of Spirit and Thought

Rapid increase of muscle movements after landing generates the sympathetic nerve and pyramidal system, so that the whole body begins to move for searching rhythm. According to the demand of the intestinal visceral system, the somatic muscle system performs rhythmic exercise. Down to human beings, rhythm has been memorized by the width of the hand, or the leg, and measurement has started. This was the beginning of number, and was that of thought and spiritual activities. Namely, it is rhythmic exercise that realizes the demand of the intestinal visceral system, and planning how to move this is the beginning of thought. Generation of thought and spirit starts like this.

Let's think of it by a protomorphic shark. One seeks the place of food and reproduction using the eye, nose, ear, and sense of touch being paraneuron. This is the beginning of "move by feeling", or impression; it is the starting point of one's heart. If one finds food, one measures the distance with the eye. This is the beginning of "thinking", and measurement is converted into the movement of a fin.

Spirit and thought have existed in our somatic muscle system. The vertebrates are created so that they can easily memorize how to move their body. Especially in mammals, this tendency is prominent, and by this memory has been established. Remembering to the extent that you can move your body unconsciously refers to memory. "Oku" refers to the world without both comfort and discomfort, or reflex movement. Namely, it is the unconscious state in which the body moves by the pyramidal muscle system. You can't remember languages, if you don't use the tongue, voice, and the mouth repeatedly in loud voice actually. If you want to remember Chinese characters or spelling, you have to write and practice. All muscles constituting the upper arm and hand remember rhythm and curve drawn by the hands as the whole. It is a performer of instrumental music who can relate tone, tune, and rhythm with muscle rhythm, position or interval of a musical instrument. A man who is good at converting tone of colour, and rhythm of line and space into muscle movements, becomes a painter. Thus, artists such as a vocalist, composer, sculptor, wood block artist, dancer and so on, have been born. A man who is good at giving treatment for broken subjects or body by grasping the cause or law of breakage, is suited for a repairman, or a doctor.

Movable Tongue and Development of the Pyramidal System

The easy movement of the tongue is due to diminution of the branchial arch cartilage accompanied by retraction of the branchial intestine after landing. The rhythmic movement of the branchia has disappeared by becoming proficient in pulmonary respiration; the branchial arch cartilage has retracted from the tongue being an aggregate of the branchial muscle and branchial arch to become a lingual bone. As a result, the tongue composed of the branchial intestinal muscle released from the skeleton begins to move. Since this coincides with the development of the sympathetic nerve and the pyramidal system, the tongue gives rise to rapid increase of nerve cells in the cerebral cortex motor area that develop in accordance with rhythmic movement created by will being the characteristics of the somatic stratified muscle. This is also due to use of the law of Use and Disuse.

At this point, the masticatory muscles, expression muscles, swallowing muscles, and vocal muscles derived from the branchial arch except muscles of auditory ossicle, transform themselves into the muscles to move even by the will of the somatic system.

Spirit and thought activities develop by the rhythmic movements of the muscle group of the branchial and excretory intestine system; the functions of the somatic motor system and pyramidalk system have been stratified into the branchial and excretory intestine supporting the basic constitution of the archetype branchial intestine such as eating, respiration, digestion, and reproduction having lived only by the parasympathetic nerve-pyramidal system.

"Language", or rhythmic movement in harmony with respiration acquired in cooperation with the facial expression muscle, tongue muscle and cervical muscles groups in the part of the branchial intestine, has developed spirit-thought activities rapidly to give gise to human beings. Language is what rhythmic movement of the muscle groups functioning in peristalsis of the visceral cranium being eating-mastication and in respiration has been diverted to the rhythm in the sound level of thought-expression of the sympathetic and pyramidal system. The infants who failed to acquire suckling at the early stage, can't often speak or think, even if the paediatricians find no abnormal observations. If you fail to acquire a series of peristalsis at the portion of the tongue, cheek, and throat caused by suckling at a certain early stage, neurons (nerve cells) of the brain can't develop well so that not only language but also ability of thought does not develop. The movement of the branchial movement is essential for the development, differentiation, and functional development.

Bushido (Chivalry) And The Gut Culture Of The Gut

At the introduction of Western civilization welcoming Meiji restoration of westernization, it is Japanese spirit combined with Western learning that was often hoped. Since the soul refers to the essential heart, it is an idea to import and assimilate Western social systems produced by the Western spirit into our culture by the Japanese heart.

Inazou Nitobe representing the wisdom of Japan in Meiji era, examined the causes why our people not having received religious education at all had the behavioural precepts based on high ethical standards, and found out its source in "Bushido" (chivalry). A Belgian lawyer, de Laveleye once asked Nitobe that Japanese has religious education or not. He answered no; it originates in "then, how religion education is taught", and it was also the answer for his wife, Marry Peterson, Elkington Nitobe's question (married Nitobe in 1891) "Why are there these ways of thinking or customs?

Inazou Nitobe stated that the heart or soul resides in the gut quoting from the Old Testament in his book "Bushido: The soul of Japan, Disclosure of Japanese thought", and showed the validity of hara-kiri. He was a Christian thought committing suicide to be banned, but he praised warriors' disposal of themselves by hara-kiri as the noble manner requiring self-control accompanied with the highest mentality. Hara-kiri originates in the manner that proud warriors do with themselves at the failure of self-realization. They knew that self, or the main body of self, dwells in the gut of the intestinal visceral system, and all desires rise from the gut, or the root of life.

At the gut of the abdomen, the information that reached the visceral brain does not come up to the level of consciousness as one's emotions. The gut in the abdomen is quite pitch-dark. There are lots of the ancient Japanese languages pointing to the truth so that they were conceivably produced by the visceral brain of the limbic system. Nothing tells the truth more phylogenically than the ancient Japanese languages of the relationship between life and the gut. It has been believed in our country from the ancient times the soul and heart dwell in the gut and the source of self lies in the gut. In Japan, there is even a verb called "haramu" of the gut meaning carrying a new life.

Since the embryos of all animals have the same shape in our country, Magatama (comma-shaped beads) symbolizing the soul of life dwelling in the gut have become a charm of Our Imperial Ancestors (one of Three Secret Treasures). It is in the uterus of the gut where the soul dwells that Magatama reside.

When viable hopes suffer a setback, the internal organs, especially the intestines and the heart ache. This is because the source of five desires lies in the gut. A military commander in our country performs hara-kiri when he realized that his self-realization had failed, because the internal organs and the bowels keep seething with rage without hara-kiri. In case the anger does not especially calm down, he does not make anyone assist him in committing hara-kiri by beheading him. The anger calms down amid the heartbreaking agony, while the intestines live for up to 8 hours after cutting the gut up. When you are vexed, your gut becomes boiling with rage, and when you are taken in, you cannot contain your anger.

That hara-kiri loses even a trace of irrationality in our mind, is not only due to the association with the foreign. The work of this portion especially chosen in the body is based on the anatomical belief in terms of the loci of the soul and emotion. Moses wrote to Joseph, "the gut loves its brother"; David prayed to God, "Don't forget his gut"; Isajah, Jeremiah, and ancient others communed with spirit, told "the gut sounds or aches"; they had all asserted that the soul dwelled in the gut (the belief having prevailed among the Japanese). The Semitic told out of habit the fat around the liver, kidney and surroundings was the loci of emotion and life. The word, "the gut" has broader meaning than Greek's phren or tumos, and both the Japanese and Greek had thought that the soul dwelled somewhere around the gut. Such way of thinking has never been defined to ancient nations. The French have asserted that they use the word, ventre, in the significant meaning in terms of physiology, though it is so vague in terms of anatomy, despite that Descartes , one of the greatest philosopher, proposed that the soul dwelled in the pineal gland. Similarly, the word, entrailles means affection and sympathy in French. (...)

The readers now understand that hara-kiri was not a mere act of committing suicide. Hara-kiri was a legal and ceremonial system. Hara-kiri generated in the Medieval Period, was an act by which the warrior expiated his sin, apologized for his mistake, evaded shame, sacrificed himself to save his friend, and verified his sincerity. When hara-kiri was executed compulsorily as the legal punishment, it was performed in a proper ritual. It was refinement of committing suicide, and nobody could not perform without extreme coolness of mind and behaviour. By these reasons, hara-kiri was especially worthy of warrior's duty. (Bushido, Inazou Nitobe; translated by Masahiro Sato, Kyobunnkann).

The other words concerning the gut (hara); 'hara ga tatsu' means 'beside oneself with anger', 'hara ise' means 'out of spite', 'hara wo watte hanasu' means 'have a heart-to-heart talk', 'hara guroi' means 'black-hearted', 'harade ugokasu' means 'moving people by the gut', 'hara wo saguru' means 'proving a person's real intentions', 'hara gei' means 'a knack for making one's views felt' and so on. Hara is the word expressing one's heart, mind, and will, which is not used in the good meaning. The reason is because it is something involving self accompanied by one's emotions.

Hara is something concerning self and emotions, and self-realization is the reproduction on an enlarged scale of spiritual self. The reproduction on an enlarged scale of self body is reproduction, which dwells in the gut (hara). Reproduction should be performed by hara, and the source of self-realization, or self lies in hara. "Bushido; The soul of Japan" by Nitobe clarified that hara-kiri was the manner of warriors peculiar to our country to do with themselves, and was based on the unshakable faith that the soul dwelled in the gut.

Culture Of The Brain

The West is called today as the culture of the brain. But, it goes without saying the descriptions of the Bible that in old days visceral brain thought in which mind dwelled in the gut or heart, was popular. Visceral brain thought refers to thought by the limbic system, which grasps the circumstances essentially as it was, and in which ancient people were versed.

A man of the present age has almost lost deep visceral brain capability because of no thinking deeply enough by the cerebral cortex. The word, guts has the meaning of courage, will-power, and impudence in addition to visceral organs, and intestines; the word, visceral, means the emotions such as emotional, on rude instinct or irrational in addition to the meaning relating to the viscera.

Christian world for two thousand years has been too long. Human bodies have all been considered to be spiritual phenomena by believing the virgin birth that was not admitted by Judaism or the Resurrection of

Christ, the Son of Man. No one has thought of the place where the soul dwelled or the virgin birth occurred. And at the age of Descartes, no-thinking deeply enough by the cerebral cortex has become active. On the other hand, in Japan 'the soul of language' thought in the ancient age has been succeeded without a break from the age of Kojiki to Edo period.

The Japanese Spirit and Magatama (a Comma-Shaped Bead)

Nitobe says in his book 'Bushido' that if religion is no more than 'moral law influenced by emotions' (Arnold Mathew), there exists no ethical system certifying religion than Bushido (the code of the samurai).

Norinaga Motoori sang expressing the silent sayings of the nation as follows.

Asking the Japanese spirit in tanka poetry,
Wild cherry blossoms smelling tomorrow.

When Shouinn Yoshida were arrested and sent to Edo by having failed to navigate with Perry, he sang the following poetry before the tombs of the forty-seven Ronin at Sengakuji temple.

Knowing that such conducts lead to such results,
Can Not Help Displaying the Japanese Spirit.

Motoori was a paediatrician, who had restored ancient morality and had accomplished the study of ancient Japanese thought and culture. He lived in Matsuzaka, who worked under Kishuu Tokugawa clan and wrote 'Tamakatsuma'. 'Katsuma' refers to a palanquin or a keepsake. He had offered the theory of politics to his feudal load as the keepsake of the spirit. On the other hand, he had remained 'Tamakushige' as literary criticism. 'Kushige' refers to a case of a comb. Viewed from the thought of the soul of language, kushige leads to kishige (strange) so that 'Tamakishige' means "strangeness of the soul". Both a case and a box are the place in which the soul dwells. Norinaga's theory of literature was stored in the box being a work of art and craft, and his theory of administration was stored in the cage of usual utensils; These were the sign that he had acquired thoroughly the theory of the soul of language.

There are valuable things called Magatama (a comma-shaped bead) that have deep relation with the soul. They are ancient ornaments, a large number of which have been discovered at the ruins of Japan and in the Korean peninsula. Since the symbol of succession to the Imperial

Throne is 'Yasakaninomagatama', Magatama must have carried something that can be recognized only by our phyla (Magatama, Yuu Mizuno, Gakuseisha). What symbol is Magatama?

Magatama has been peculiar to our country, and have many forms viewed from all aspects. A certain Magatama is of the branchial intestinal germ having niches corresponding to the mouth and branchia, which is Choujigashira Magatama. A Magatama having the umbilical cord has been unearthed. There is also a Magatama at the stage of the nerve germ, which is Komochimagatama (a comma-shaped bead having a baby). In Korean peninsula, a treasure crown with a hanging golden fish along with Magatama, has been unearthed. What are common among them having various shapes, are that Magatama having golden cap in the face and head have been unearthed, all of which have moulded the shape of an embryo, while Magatamas having the grooves of the eyes, mouth and branchia that only embryos have, and ones having the umbilical cord, are noticeable.

A fish and an embryo are the symbol of the life of vertebrates in phylogeny. Magatama has been considered to be one of the things succeeded by the head of the race from generation to generation, which was a charm of traditional treasures to transmit the life of the race for ever, taking care of Magatama moulding an embryo as the symbol of life. Why had ancient people realized the ancestor of life in a long time ago? It is quite easy for our race in whom an equestrian people and a marine people were mixed. Comparing the embryos of the edible animals from a cow, horse, deer, rabbit, bird, fish, giant shark, to a whale, the shape of the embryo which Haeckel had seen, had appeared as large as just a bean as the common shape. This might have been designated as Magatama that is the symbol of life common to every life and soul. This is the ground for the guardian deity of the race loving the soul of language. The far source of the intuition may depend on life memory based on the cellular level.

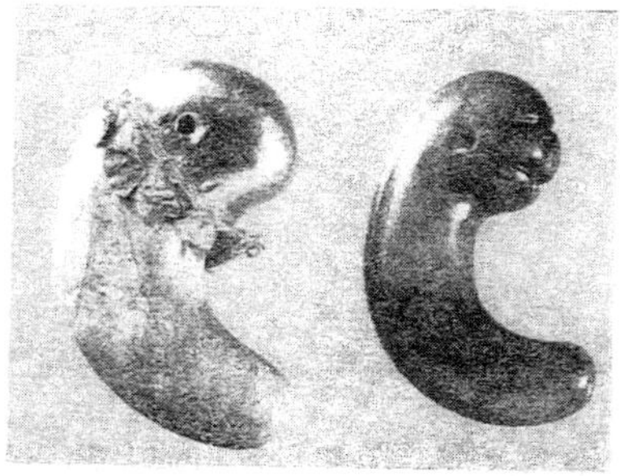

Fig. V-3. Magatama (Comma-shaped beads)

In the soul of language-thought, Maga in Magatama means "true or real". Tama means "soul", that is, "life". Namely, Magatama means real life. In olden days, the sound of Magatama which was produced by touching each other, was called "Tamayura", which was designated to be the symbol of the soul's touching. You will find the meaning and sound of the language suitable for the country in which the soul of language was coming and going. Magatama at the initial stage has been said to have been produced by a hard bead yielded around Itoi River in Niigata Prefecture. Magatama that was a treasure in those days, has been discovered not only in the relics of Japan, but also of Korean Peninsula, and many of them had been made in Japan.

The origin of Magatama has been discussed in many ways such as the theory derived from 'a tusk of the animal', the one of an embryo, and the one being the symbol of the moon. To sum up, there are the theory of fish-shape origin, the one of the kidney-imitation, the one of an embryo-imitation, the one of fish hook origin, the one of the animal tusk origin and so on (Magatama, Yuu Mizuno, Gakuseisha). To be sure, the tusk-Magatama (a comma-shaped bead like a pierced tusk) has been found in China and Japan at the beginning of the Jomon period in the Old Stone Age, and comma-shaped beads have certainly been a guarding deity for the Imperial ancestors as a symbol of an embryo. It has been written in Kojiki, the legendary stories of old Japan, that Yamato Takerunomikoto had got through a crisis by means of a comma-shaped bead made from flint. Phylogeny backs up strongly Haeckel's Recapitulation theory. This Recapitulation Theory asserts that a primordial head, 'Kaput' of a vertebrate repeats itself. Looking at the comma-shaped bead with its head covered by gold, we acknowledge that it is just the prototype of life. As previously mentioned, comma-shaped beads unearthed in our country or the Korean peninsula include those with heads bearing golden decorations, those with gill-slits inscribed in their heads, and those having roe with the umbilical cord of the nerve germ shape of a sea horse type. What's more, around the crown or necklaces having been unearthed in the Korean peninsula, fish made from golden board along with Magatama has been hanged. Ancient people might have known intuitively that our distant ancestors were vertebrates, and were fish living in the sea in the myth era earlier still. A myth might be a story of the genealogy of the world carried in life memory. Magatama, by the way, has been unearthed in Japan, the Korean peninsula, and Egypt.

Soul of the Samurai and Soul of the Mammal

Incidentally, the spirit of samurai has been said to be "sword". To view the vertebrate from the standpoint of this logic, the spirit of the mammal especially among these phyla turns out to be the tooth. Loss of teeth results in the death of the many mammals except a man.

The definition of the mammal is "the vertebrate born having the suckling system that is to masticate". Suckling refers to lactation, which means suckling milk. The trait of the mammal is having Heterodontia, and the most important trait is having three kinds of teeth. This is the definition of the mammal. The shape of the tooth differs according to the site of the jaw, because an articulator (the periodontal membrane) between the bone and tooth exists. In the

Chapter 3, I described in detail that there was a shark called Heterodontus japonicus. This refers to "Japanese shark having mammalian type tooth". The Japanese name of it is Nekozame. In fact, the face of adult organism of this shark corresponds quite with the parts of the human embryo. This has verified that Heterodontas laponicus has become the mammalian type reptile, as Recapitulation theory by Haeckel has shown. The egg of this Heterodontus Japonicus is larger than that of a chicken, and is composed of a kelp-type bag. The cellulose at the base of an ascidian has been succeeded to the egg shell of a shark. The young fish just before the egg's hatching is quite similar to an embryo in the 32nd day both in size and shape. Ancient people must have observed earnestly the egg of a shark, an embryo in the egg of a bird , and an embryo of a rabbit in order to produce Magatama with spirit.

Harakiri cutting up one's own belly by one's will in which the soul resides would be impossible to carry out without the world famous Japanese sword. The sword is also one of the three Sacred Treasures of the Imperial Ancestors as well as Magatama. It is Amenomurakumonotsurugi.

What the soul of a warrior is to a sword, the modesty of the code of the samurai is to readiness for hara-kiri by servants (Hagakure). The figure of life is Magatama moulding the soul. It is Yata's mirror that moulds the sun which has become the symbol of the heart and the origin of everything. Three sacred Treasures have been succeeded profoundly and secretly to the Imperial Family with the transmission of Buddhism since the age of the Imperial Ancestors, Kamuyamatoiwarebiko. They have been succeeded as the symbol together with the code of samurai, the soul, and the warrior's attitude to behave. It requires a strong spirit to prove one's innocence or seethe one's anger by cutting up the belly in which the soul resides with the sword of samurai's soul. Severe pain accompanies self-death. It is regeneration of one's own, or reproduction that life is heading for. Reproduction is the goal of life, and death is the completion of life. The origin of life lies in the gut, and reproduction also has its origin in the gut, which is carried out by the guts. According to "Hagakure no Eichi" (Sagacity of Hagakure) by Yoshiaki Koike (Koudansha Gendai Shinsho), the fall of a castle had involved hara-kiri without fail, and before the fall of a castle the sight of sexual intercourse had been seen everywhere. At Tokyo air raid of the Second World War, a playwright had encountered odd sights in the compound of Yasukuni shrine around the time at which the colour of flames at the sky had thinned. Though it was still half-lighted in the neighbourhood toward evening, harsh sexual intercourse has been carried out in the shade of bushes. They had all worn air defence hood, air defence clothes, or women's work pants. Their conducts were far from sensual pleasure. Their acts were not felt to be sensual, and were temporary and evasive sexual acts. Nobody paid lewd attention in the neighbourhood. It was the fact in broad daylight just before the fall of the Japanese Empire. Reproduction being the goal of living and death by hara-kiri or by air raid, are two sides of the same coin.

After Word
My Research Activities by a Frontal Breakthrough)

Research into the Vertebrates that I Often Wanted even in Dreams

As previously mentioned, I had the pleasure of meeting Prof. Shigeo Miki at his lecture of 'Morphology of Life', and awakened to what research is. I entered the biochemistry department office of Tokyo University, and obtained a degree in molecular biology concerning mitochondria at the cutting edge of those days about 40 years ago.

"Desire to study the evolution of the vertebrate" once in a lifetime, was the reason why I returned to the postgraduate course forty years ago. The enigma of evolution is most difficult, and no scholar has studied demonstratively as a science through 19th to 20th century. I've thought of studying the most difficult problem once directly. So I decided to start from what I could do at the age of 40. Since I was a clinician, I wanted to become a doctor of oral medicine the importance of whom Tohru Shimamine had advocated. I've changed the clinical name of "Department of dental and oral surgery" to "Department of oral medicine", and aimed to become a superior oral doctor, or the doctor out of the doctors (a superior doctor).

The most difficult theme in the department of oral medicine is to regenerate the tooth artificially. Especially, it is almost impossible to regenerate the whole tooth by starting from the tooth germ, since it takes a long time and control of the figure is difficult. However, all we have to do is the following procedure: First of all, we produce simply the tooth crown and root, and implant them by operation. And we induce by any means the vital tissues joining with the body, or the periodontal membrane accompanied by cementoblasts, and the lamellated bone supporting the periodontal membrane.

Since the age of 40, I have followed this way of thinking, and for some five ears I've exercised on desk plans. It is central-breakthrough-thought that grapple directly with the most difficult problem in the department of oral medicine. Five years later, I've succeeded in producing artificial dental roots; It took around 10 years' clinical research in order to apply them to men without trouble.

During those periods, many problems were caused. Man may feel jealousy. The most intimate friends or superiors had tried to dash the artificial roots vehemently. But owing to the resistance by the opponents, my research was performed very carefully, and a clue for discovery was obtained; I could discover that Lamarck's Use and Disuse theory was dependent on gene expression of mescenchymal cells by streaming potential caused by biomechanics on the way to the study of biomechanics concerning gravitation. I discovered the factor inducing cells composing the cementum or periodontal membrane which no one has not found out; It was energy of repeated load having 0.2 mm amplitude (repeated fluctuation loading by mastication and occlusal movement). It has been suggested that the evolution of the vertebrate has occurred by gravitational energy of biomechanics. It was since 1988 that I properly grappled with the biomechanics concerning gravitation, which was a clue of becoming aware of a loophole in modern medicine. In those days, hydroxyapatite which is the characteristic mineral distinguishing the vertebrate from others, has been synthesized, and sintered ceramics has been put to practical use. The author has succeeded for the first time in the world in developing artificial roots being an alternative artificial organ for the tooth.

At a glance of a specimen of a dog's mandible in which artificial roots have been implanted with syndesmosis and alveolar bone proper formed, I found out that the teeth and the bone of mandible had structures responded to dynamics. In order to verify this mathematically, the cutting edge dynamic analysis (finite element analysis) was employed using this specimen as a model, which was the beginning of biomechanical research.

At the end of this year, I have established AART Association (Apatite Artificial Roots Therapy Association). I had learned the basics of mechanical analytic methods from Professor Shigeru Nakagiri in the Institute of Industrial Science in order to analyze the dynamics of artificial roots, and presented the research results at the first WCB (World Congress of Biomechanics-begun by Professor Fung) held at California University of San Diego in 1990. I remember that I was more than a little strained when I had read it under the chairman of

Professor Brunski, a successor to Paulac who discovered streaming potential. At the aquarium of Scripps Institute of Oceanography at La Jolla in San Diego where I happened to drop in on, I had seen the living animals of each stage lined according to the stream of phylogeny; later, it became the clue to develop an experimental evolutionary method in order to elucidate the enigma of the development of bone marrow haemopoiesis.

I realized intuitively that three enigmas of the vertebrates, 'modification of morphology and function', 'development of the immune system', 'development of bone marrow hemopoiesis', were systematically lined up as the responses to gravity, inertia, and active-reaction-law by the usage of the body on each stage of phylogeny. I made friends with Professor Michael in Poland specializing in haematology at the congress, and talked about the development of artificial bone marrow hemopoiesis nest by biomechanical energy. The first thing I wrestled with after having returned was application for development research of artificial bone marrow to the Ministry of Education, Culture, Sports, Science and Technology. Fortunately, this was accepted smoothly. I advanced to the development experiment of real artificial bone marrow chamber right after preliminary research, and all researches had all come off fine.

In 1994, I presented artificial roots that induce the cementum and periodontal membrane at the invitation lecture of CIMTC, an international congress of ceramics that has been held every four years in Italy, and next I could present the compilation of the therapeutic methods for deformities of the face at the second WCB in Amsterdam. This year I received the Original Top Prize for the development of artificial bone marrow at the Japanese Society of Artificial Organ. Now, finally you can start the research of the vertebrate.

Tracing back the origin of biomechanics, I knew that Wilhelm Roux had originated this science by admitting 'gravitational function is essentially important for the development and evolution of life', together with Die Entwicklungsmeckanik at the early stage of this century. He clarified also the relationship between ontogeny and phylogeny, and was the first disciple of Haeckel who had established biogenic law (recapitulation theory).

Since I came from Yokosuka city of Kanagawa prefecture, I was introduced Hiroshi Kabasawa, the director of Aburatsubo Marine Park Aquarium at the time of alumni society meeting of the junior high school. I started there Experimental Evolutionary Study that induce haemopoietic nests artificially in artificial bone marrow by implanting artificial bone marrow made from ceramics into sharks having no bone marrow haemopoietic nests. This study is the anticipating one of evolution to induce bone marrow nests in the animal of one stage before in evolution. All these studies came off fine. In the next place, I started implantation-research of the shark, eft, Xenopus, and mammals.

I bred many animals such as an eft, mouse, Ambystoma, rat, or quail in my house, and carried them to Miura peninsula to operate, and carried them back to home to give them postoperative management. I used to go to a pet shop in a department store to prepare for useful utensils for postoperative management and breeding, and my wife alone took care of them, and it was very troublesome. Such difficult researches have come off very fine owing to my wife's devoted cooperation for which I was very thankful. We took care of rats or quails late at night, and before long we bought Russian blue (a cat) at the department store. It was 10 ㎝ in size and let out a meow feebly.

My research has started from Linne, followed by Lamarck, Goethe, Haeckel, Roux, Miki and Nishihara; I reached an agreement with my wife that we should breed one cat and two dogs named LiLA, HeRuG, and MN, taking the head letter of those researchers. First of all, this Russian blue was named as HeRuG. Researches came off very fine and transplant-operations also succeeded; there were 20 animals in total: mice or quails in which the skin of a shark was grafted, and rats in which a part-brain of a shark was transplanted. We bred rats at the entrance hall, and they had returned to the cage in the morning after having taken a walk at midnight. Since the hall for breeding became narrow, we made an experimental pen in the garden. It was in this pen that we made experiments to breed Xenopus and axolotl on the land.

I collected the researches and published "Gravity made living organisms evolve" in Bluebacks. I was beside myself with joy, since I grasped the clue to break through the most difficult problem in life science. Anyway, I vanquished Darwin's theory and showed for the first time the alternative scientific idea. My research was introduced in the program of 'Interesting study life' and 'Science Eye' of NHK, and also introduced at Chyukyo TV; I was obliged to go to the Marine Park of Aburatsubo devoting one day making a film many times. I was perplexed because the animal pen was very often photographed at night. Therefore, I decided to rebuild my house. At the turret of the rooftop I made a breeding room for experiments, but in those days, economic conditions in our country have greatly changed; all the shops selling Axolotl or Xenopus in the inside of Yamanote Line in Tokyo were wiped out. Five years ago there lived 20 animals, but now I breed one

Axolotl of some 20 cm and one Xenopus of 8 cm alone. Meanwhile two toy poodles were added to the family, and LiLA, HeRuG, and MN were completed. They are by-products of the research of evolution. Although the evolutionary pets are hardly worth the names, by having observed LiLA, HeRuG, and MN, I realized that both a dog and a cat (the vertebrate) were nearly as clever and full of more feelings as man. And having observed rats, LiLA, HeRuG and MN, I could understand well the way to acquire respiration by the baby of the mammal.

Frontal Breakthrough

Since I've learned 'General Morphology' at the age of a student, I've taken it to heart that science was to disclose by the law of causality the law behind the perplexing phenomena on the basis of natural observation; I have come to think that the theory of evolution or views on pathogenesis that were teleological way of thinking of life science having been talked by the conventional medicine or anatomy, was wrong. I was hardly interested in the already completed operation methods in clinical medicine. Since the operation of harelips or cleft palate in my region has especially been accomplished, I wanted to do with several cases, but I have not have the slightest intention of making it my life-work. In the medicine of surgery, it is the best way of thinking how to do away with operations, and it is the second-class surgeon that is proud of having experienced several hundred cases' operations developed by foreign surgeons. In that sense, there are few first-rate or blue-ribbon medical scholars. The medical world after having been defeated in war, has pleased to get a tiny share of NIH (National Institute of Health) research in U.S.A.

I returned to the road as a scientist at the age of 26, and graduated from the post graduate course at the age of 30. Immediately I became an assistant professor; the theme of the research was the wide problem that was not solved yet in those days, or the most difficult problem that was to be solved by the different methods from one of the trends of the times. Accordingly, immediate results were not expected at all in every research involved.

The first theme was as follows: Why does a handsome young man or young woman have a transformed face and have a curvature of the spine as he or she gets older and gets ill? I can't say why, but I was convinced that periodontal disease or loss of teeth by caries was a kind of deformity. About ten years later after having become a doctor of oral medicine, I realized roughly that a causal factor of deformities was in 'the oral cavity and its peripheral habits'; to sum it up biomechanically, I had to solve the relationship between generation and remodelling of bone and biomechanics after having succeeded in developing artificial roots having been the second theme. The developmental research of artificial roots that was not yet succeeded in the world, was also due to a frontal breakthrough. The most difficult problem in the department of the oral cavity can be surprisingly simply solved by grappling with just in front. It requires a frontal breakthrough idea of grasping the essence of the vertebrate, the most difficult problem in life science. As the defining substance of this phylon is bone, artificial roots and artificial bone marrow could be developed by any contrivance, if mineral substances of bone are artificially synthesized.

However, the development of artificial roots and bone marrow that is the most difficult problem for human beings, would be impossible without improving the human relationship among the scientists in the department of oral surgery of Tokyo University. That was the system in Japan. The author was not in a good environment as to my boss; I was opposed two times to the head of the department, and I broke through the difficulties two times. Therefore, I could live a free research life.

Shortly after I began to research biomechanics on artificial bone marrow, I attended a meeting for the research on Priority Areas by the scientific research funds under the Ministry of Education, Science, and Culture. At the meeting I presented with enthusiasm; I disclosed by the research on biomechanics of bone that mechanical stimuli have been converted into hydrodynamics of the body fluid, furthermore being converted to streaming potential, and as a result, the trigger of the genes of the mesenchymal cells has been pulled, and BMP (Bone Morphogenic Protein) has been produced. The professor of a certain university, the chairman asked me "Who has said it?". When I answered, "It's me", my opinion was at once dismissed. Science is absent from the Japanese professional scientists: only the views by foreign scientists are significant in our country. In this way, Japan is always subject to be defeated in the field of science in the world. At last, I began research on the disclosure of the law of evolution of the vertebrate, the disclosure of the immune system, and the disclosure of the causation of the development of immune diseases that human beings could not accomplish for 100 years of the 20th century. The frontal breakthrough is just indispensable for this. It was about at the year of 1996 that I got around to full-scale work.

Since I established Oromaxillofacial Biomechanics Congress several years ago, I received a professor emeritus's guidance on biomaterials. He advised me then; what you want to do is so difficult that it would take 30 years for the chairman of the department of 30 years old having 30 followers to accomplish, so you should stop doing it. The most difficult problem is usually avoided solving because of the least success. However, no matter how difficult it is, there should be a vital point to be solved.

The greatest riddle of the vertebrate is evolution. I have gotten a general idea that the vital point of evolution must lie in the landing drama, the second revolution of the vertebrate which was the sudden change in the course of evolution. The close observation of it must lead to the solution of the riddle. There were no scientists or doctors who had observed accurately this landing drama in evolution. It also requires strategy to elucidate science. The most effective strategy is a frontal breakthrough-thought since olden times. It is adapted especially for science. Every one of them, Pasteur or Koch who originated bacteriology, Goethe who created morphology, Lamarck who elucidated evolution, Haeckel who created biogenic law, and Cuvier who established the system of palaeontology and comparative anatomy, explored and created the science that had not yet existed. The researches carried out by the team composed of a number of persons are all nothing new and are mass-produced; namely, their jobs are for living. As a matter of course, there is no need for originality. Medicine and life science after war in Japan were almost in such conditions. They could not come up with. In Japan after the war, there are few life scientists having thought of a frontal breakthrough, and there are remarkably many occupational scientists.

If you observe well the process of evolution, you will naturally notice by yourself that the key to solve the law of the transformation of morphology lies in the landing drama. There have been by now few scientists who observed correctly the change in that time and its causes materialistically and biologically. Shigeo Miki had observed to a considerable degree, but had failed to notice too much degree. As far as immunology and the investigation of the cause of the development of immune diseases are concerned so far, even the comparative observation of the difference between the patient and the normal person has completely been missed. Under those circumstances, both evolution and plagues have become unknown happenings, and therefore the causal factor of evolution or the riddle of the development of immune diseases can not be solved. Namely, the conventional investigating method of evolution or the development of immune diseases has neglected the observation of nature being the basics of the solving methods of riddles far from a frontal breakthrough. I introduced the law of the conservation of energy into the observation of nature that had been the greatest achievement in science in the 20th century, but had been completely forgotten, and broke through from the view point. That's why I could solve the difficult problem in the major portion by myself. This has made it possible to obtain the clue to elucidate the developmental science of mind and spirit.

I have thought for a long time that the job to create a new science by solving the riddle of the vertebrate was compared to the battle capturing an impregnable castle. In order to do that, it is necessary to reconnoitre the enemy position, know its weak points, and build up the strategy after setting the goal. The frontal breakthrough capturing the enemy's headquarters that is considered to the most difficult to attack, is the most effective method capable of success by a minority from the ancient times. The author has used this method in everything.

One day, I described the method by a frontal breakthrough to solve the riddle of evolution and the immune system by developing artificial bone marrow with a hybrid method in the paper of Japanese Journal of Artificial Organs; as the term 'a frontal breakthrough' is the word for military affairs, and is not a scientific word for life science, I was told to change the word. Since it was quite reasonable, I changed it into solution for a hard problem.

Results of Research and Mass Communication

The theme in my research was too wide in range, and essentials; it handled only the most difficult problem and it seemed to have gone against the stream of the times; I became aware that since the starting point in research, my research had a bad reputation among occupational scientists. My presentations in our country continued to be ignored.

With the presentation of artificial roots at Seoul in 1989 as a start, I shifted the presentations at the congress into the international congress such as Biomechanics congress, congress of Bio ceramics or Bioengineering, and congress on Ceramics in Medicine. And I have presented at the international congress once or twice in a year at the following cities and countries: Drameheler (Canada), San Diego, Terre Haute (USA), London, Chester, Berlin, Chicago, Filenze, Amsterdam, Finland, Sweden, Istanbul (Turkey), Beijing, Taipei, Bologna, and Faenza (Italy). However, It was not until 1994 that the results of research were given much publicity by the press. Before being given much publicity, I was strolling around the port of Stockholm with my wife between whiles of the congress in Sweden. Suddenly, our pictures were snapped one after another. It was just the time to get on a small boat after having looked around the government office building holding the awards ceremony for the Nobel Prize. When we returned to the hotel after sight-seeing the next day, an information clerk of the hotel rushed upon us, and to our surprise brought the newspaper illustrated in a big way with photograph of us at the front page of Stockholm News. After I returned to Japan, 'oral breathing' was taken up in Tokyo Newspaper, and after a while, 'artificial bone marrow' was taken up on a large scale in Asahi Newspaper. In a little while, surprisingly I was asked to submit the English paper from the Swedish Embassy.

A newspaper " DN.Stockholm" in which the author and his wife were introduced

I have clarified the evolution of the vertebrate responded to mechanics for 500 million years, the whereabouts of mind, and elucidated the riddle of life science by a frontal breakthrough. Let's find out the standard of new life science based on the law of the new universal constitution suitable for the 21st century. It's time to fill a new leather bag with new wine.

It was very hard to write this book: if you don't thoroughly understand and make your own 'the law of the conservation of energy' by your body, you will not understand that mind is life energy. For this, you have to think of the principle theory of relativity by Einstein by consulting the law of the conservation of energy, and you have to correct the contradictory points. You have to understand collectively under the law of the conservation of energy the following phenomena and facts: superconductivity phenomena (electrons at ultralow temperature [-273℃] speed up 20 million times faster than at a normal temperature) discovered by Kamerlingh Onnes in 1911, and the experiment data of the speed of light at ultracold temperature (near absolute zero) [17m/sec at -273℃, and 0m/sec at –273.13℃] carried out for three years at Harvard University. The speed of light is not constant. These three different phenomena are also the different sides of the same phenomena as well as the three riddles of the vertebrate. That electrons run fast at ultralow temperature and the speed of light becomes slow, means that time at a normal temperature gets prolonged by 20 million times in ultralow temperature. This is the essence of the Principle Theory of Relativity that space and time are in the relative relationship through light. Both time and space are energy.

I have succeeded in inducing bone marrow hemopoietic cells, osteoblasts, cementoblasts, or fibroblasts around ceramics by gene expression of the cells of the animal in which artificial organs were transplanted, by energy called streaming potential based on gravitational energy with the development of hybrid-type artificial bone marrow and artificial roots. As a result, I discovered that the evolution of the vertebrates has occurred by gravity, and elucidated what gravity is; the results are elucidation of three riddles and the genuine interpretation of the Principal Theory of Relativity. This is the fruit of my clarification that life phenomena are electric phenomena in water-soluble colloid organisms.

It is the Genuine Principle Theory of Relativity that space and time are in the relative relationship through the energy of light, and both space and time turn out to be energy. The speed of light is variable on thermal energy. Multiplying the speed of light by time makes always constant. This is the size of space of the place (energy state), which is always constant. It is due to the law of the Conservation of Energy. However, as Newton showed, it functions on only the substance with mass, and does not function on light, space or time at all. Since these things have been confused, everything was thrown into disorder.

If only the reasons are understood, 'mind' and 'spirit' and 'thought' that are the riddles of medicine, or immune diseases in life science are easily elucidated. 'Mind' or 'spirit' is sustained by energy and energy metabolism, and their modulation leads to the development of immune diseases. No matter how you talk tediously as far as medicine is concerned, there is no point if diseases can not be cured.

Today's medicine in our country is near to the conditions at the times of diagnostic nihilism proposed by Skoda of the Vienna school of the 19th century. Since Skoda insisted that diagnostics was more important than life of the patient and competed for pathological diagnosis at autopsies, he was severely criticized in those days. Correcting the wrong intake of energy by controlling energy and its metabolism, leads easily to cure intractable diseases, because life is electric phenomena of water-soluble colloid in outer space. Therefore, both mind and qigong can be captured as light by the camera. The establishment of a new life philosophy lies on the hands of the Japanese. Let's endeavour to achieve self-realization by knowing energy well.

The research was supported by the following Grant-in-Aids for scientific research from the Ministry of Education, Science and Culture, Japan.

1) [Research on the development of artificial bone marrow] 1991-1993, A test research (B) 03557107. a report
2) [Elucidation of the mechanism of morphological functional adaptation of the bone-biomechanics of the bone and comprehensive study of piezoelectric quality], 1993, Research on Priority Area, (1) 05221102, a report.
3) [Elucidation of the mechanism of morphological functional adaptation of the bone-relationship between biomechanics of the bone, and bioelectric current, and physiologic active substance], 1994, Research on Priority Area, (1) 06213102, a report.
4) [Development of artificial bone made of sintered hydroxyapatite-collagen composite], 1994-1996, Basic Research, (B) 06558119, a report.
5) [Analysis of deformities of the maxillofacial morphology due to environmental factors], 1994-1996, General Research, (B) 06455008, a report.
6) [Development and practical use of artificial bone marrow-Practical use of hybrid-type immune organ · the inducing system of artificial bone marrow hemopoietic nests], 1995-1997, Basic Research, (A), (1), 07309003, a report.
7) [Exploration of a new evolutionary theory through experiments – Establishment of the experimental system of the evolutionary study responded to biomechanics in the vertebrates], 1996-1997, Priority Area, (1) 08233102, a report.
8) [Research on the development and practical use of artificial bone marrow, and the establishment of a new concept of immunology], 1997-2000, Basic Research, (A), (1), 09309003, a report (2004).

Part of the contents of this book has been drastically added to the contents carried in the following magazines: [shizenshoku nyuusu (News on natural food)], [seimei to enerugi-(Life and energy), published by Shizenshoku news co.

[Yakyoku(pharmacy)], and [atarashii menekigaku (new immunology) (2002), published by Nanzandou co.

About The Author

Katsunari Nishihara was born in 1940 at Kanagawa Prefecture. He graduated from Tokyo Medical and Dental University. He completed the doctor's course in the Faculty of Medicine at The University of Tokyo-Graduate School (Ph.D.). After Assistant Professor in the department of Oral Surgery, Faculty of Medicine, at The University of Tokyo, he is now President of The Japanese Association of Therapeutics for Immune Diseases, and President of Nishihara Institute. He is the leading expert on the development of artificial dental roots and artificial bone marrow. He won the Original Top Prize at the 32nd Japanese Society of Artificial Organs.

Nishihara Institute (Medicine for health and beauty treatment)
Hara Building 301, 6-2-5 Roppongi, Minato-ku, Tokyo, Japan. 106-0032.
Phone: 81-3-3479-1462
Facsimile: 81-3-3479-1473

His writings:

[Kao no Kagaku (Science of The Face)], [Akachan no Shinkagaku (Evolutionary study for a baby)], [Oitsumerareta Shinkagaku (The theory of evolution driven into a corner)], {Published by Nihon Kyoubunsha}, [Seibutsu wa Jyuuryoku ga Shinka saseta (Gravity made living organisms evolve)] {Published by Kodansha}, [Menekibyou wa Kowakunai (Not afraid of immune diseases)] {Published by Dohosha, Kadokawa shoten}, [jyuuryoku taiou Shinkagaku (The Gravity Corresponding Evolutionary Theory)] {Published by Nanzando}, and a number of other books.

* 9 7 8 1 8 4 4 2 6 8 6 9 6 *

ND - #0184 - 060826 - C0 - 234/156/13 - PB - 9781844268696 - Gloss Lamination